THE PYROTECHNIST'S TREASURY

A GUIDE TO MAKING FIREWORKS AND PYROTECHNICS

FIRST RELEASED AS THE:

COMPLETE ART OF MAKING FIREWORKS.

BY

THOMAS KENTISH

This publication is for historical record only and should not be used to produce fireworks or pyrotechnics of any kind.

PREFACE.

Until within the last few years there was scarcely any work in English, that I am aware of, on the subject of Pyrotechny, worth reading, with the exception of an Article in "Brewster's Cyclopædia," by MacCulloch; and this, besides being accessible to only a few, having been published in 1830, made no mention of colours, which form the most beautiful part of the art.

In the first year of the present century a treatise was written by a Captain Jones, which has been copied, in whole or in part, into almost every work since published. The greater portion of it is absurd and impracticable, and shows that it was written by a person who undertook to teach what he had not learnt.

The first work of any real utility that came under my notice was a series of papers by "Practicus." This was soon followed by another, varied by the new chemical nomenclature.

The subject is far from being treated exhaustively in either of these works, so that I trust the reader will find in the following pages a fund of information, both in the repertory of recipes and the methods of manipulation.

To Chertier belongs the great improvement in colours. He was, as I was informed by the late Mr. Southby, who knew him personally, and who derived much information from him while in Paris, a retired French Artillery Officer, who made colours "his study and theme." His "New Researches," published in 1856, nearly thirty years after his first pamphlet, is an excellent work, that leaves little to be desired in the way of colours. Gunpowder attracted but little of his attention.

Tessier, of Paris, has written, since, his "Pyrotechnic Chemistry," and a new edition of the "Firework-Maker," by Hutstein and Websky, was published three or four years ago at Breslau.

I have neglected none of these sources of information, but do not know that I have been able to learn much from them with which I was not previously acquainted.

Gibbon's "Artillerist's Manual," and Benton's "Ordnance and Gunnery," published at New York, have furnished me with a hint or two; also Owen's "Practice of Modern Artillery," Scoffern's "Projectile Weapons of War," and his "New Resources of Warfare;" but as all these works are on Military Pyrotechny, they have not been available to any great extent. I mention their names, that anyone desirous of a knowledge of Rockets and Shells, as instruments of destruction, may know the books from which to gain the information.

It is possible that, as Chemistry advances, a few new substances may be discovered; meanwhile, nearly every shade of colour may already be produced.

The extensive use of these colours, from their beauty and variety, for stars and lancework, has very materially altered the class of fireworks, and necessitates the employment of an enormous quantity of quick-match. The preparation of this is one of the most disagreeable parts of Pyrotechny, besides demanding a great amount of manipulative skill. Most amateurs are deterred from attempting to manufacture it, and so have to content themselves with only the simplest pieces.

I have, therefore, set myself sedulously to work to devise a different plan of preparing it; and have succeeded—not till after many years, however, and numerous failures, simple as it now seems—in discovering a process of producing a splendid and perfect match, that leaves nothing to be desired; which is easy and expeditious, and does not even soil the hands. Amateurs will now find no trouble in making any pieces they may desire.

Dangerous chemicals, that might produce spontaneous combustion, have been rigidly excluded; the fullest information throughout has been furnished for preparing the mixtures with safety; the mode of projecting balloons, and the manner of constructing steelyard scales, and every kind of tool used in the art, has been added; and no pains have been spared to make the work as complete and comprehensive as possible.

In conclusion, I may say that I have had thirty-five years' experience; and there is nothing I have not tried repeatedly: the reader may, therefore, place the greatest confidence in everything recommended. He has only to imitate to succeed.

T. K.

CONTENTS.

Roman Candles:

To Make a 5/8 Roman Candle
To Make a Roman Candle Star
To Damp Stars
To Make Lac Solution
To Make Wax Solution
To Make Stearine Solution
To Make Gum Solution
To Make Dextrine Solution
To Make Paste
Roman Candle Scoops
To Charge Roman Candle Cases
To Make Touchpaper
To Make Slowmatch
To Make Quickmatch
A New Method

Rockets:

To Make a 6/8 Rocket
To Charge Rocket Cases
Rocket Stars
To Make Cut Stars
Dry Pill-box Stars
To Fill the Boxes with Dry Colour
Another Way
Bottomless Pill-boxes
Enveloped Stars
Rocket Heads

Wheel and Fixed Cases
Gerbes
Flower Pots
Port Fires and Shell Fuses

Tourbillions:

To Make a Tourbillion
Saxons
Five Pointed Stars
Squibs
Serpents
Pinwheels

Crackers:

To Make a Cracker
Leader Pipes

Maroons:

To Make a Maroon
Another Method
Gold and Silver Rains
Peacock's Plumes
Saucissons
Pearl Streamers
Blue Lights and Star Candles, or Star Lights
Prince of Wales' Feathers

Lances:

To Make a Lance
To Form a Device, or Design
To Preserve Steel Filings, or Cast-iron Borings

Shells:

To Make the Shells
Cylindrical Shells
Asteroid Rockets

Compound Fireworks:

To Fire a Girandole of 100 Rockets at once
Tools
To Construct a Steelyard Scale

Montgolfier Balloons:

To Cut the Gores
To Join the Halves Together
To Paste Two Gores Together
To Paste the Pairs Together
To Make the Last Joining
To Wire the Balloon

Winged Rockets:

To Make Winged Rockets
To Construct a Slot-Tube, or Rocket-Guide
List of Prices
Concluding Remarks

Fuses:

Roman Candle
Rocket
Bursting Fire
Starting Fire
Wheel and Fixed Case
Squib and Serpent
Pinwheel
Saxon
Five-Pointed Star
Spur-Fire, for Flower-Pots and Star-Candles
Tourbillion
Bengal Light
Blue Light
Wasp Light
Portfire
Shell Fuse
Gerbe

Tailed, Streamer, or Comet Stars, for Rockets, Shells, and Roman Candles
Oiled Tailed Stars for Rockets and Shells
Steel Stars for Rockets and Shells
Pearl Streamer
Gold Rain
Silver Rain
White or Bright Stars and Lances
Sugar Blues for Stars and Lances
Blue Stars and Lances without Sugar
Crimson and Scarlet Stars and Lances
Green or Emerald Stars and Lances
Deep and Pale Yellow Stars and Lances
Mauve and Lilac Stars and Lances
Purple and Violet Stars and Lances
Magnesium Colours for Stars and Lances
Slow Fires, to be Heaped upon a Tile in the shape of a Cone, and Lit at the Top

THE PYROTECHNIST'S TREASURY.

ROMAN CANDLES.

To Make a ⅝ Roman Candle.

Procure a straight piece of brass tube, ⅝ of an inch external diameter, and 16½ inches long. Saw or file off a piece, 1½ inch long, fig. 1. This is for the star former, and is drawn of the correct size.

In the other piece, of 15 inches, fix a handle, as shown, in diminutive, in fig. 7. This is for the case former. It should be filed smooth at the end.

Take another piece of brass tube 4½/8 or 9/16 of an inch external diameter, and about 16½ inches long. In this also fix a handle, or fix it into a handle, fig. 4. Invert it, and set it upright in a flower-pot, filled with sand or loose mould. Melt some lead in a ladle, and pour it slowly into the tube, leaving room for the air to escape up the side, till it is full. If the lead is poured in rapidly, the confined air, expanding, jerks the metal up, and may cause serious injury. A pound or more of lead will be required. When cold, drive the end of the lead in with a hammer, and file it smooth. This is for a rammer.

Take a piece of deal, fig. 6, about 12 inches long, 6 broad, and ¾ thick; and, on the top, screw a handle, like one on a copper-lid, or black-lead brush. This is for a rolling-board. An iron door-handle would serve, and may be obtained at any ironmonger's, for about 2*d.* or 3*d.* A wooden one, however, about an inch thick, not cylindrical, but slightly flat, and rounded at the edges, is preferable, as it gives more purchase for the hand.

Cut a piece of tin, or zinc, or thin board, into the shape of fig. 8, in which the distance between the arms a and b, across the dotted line, shall be ⅞ of an inch. This is for a gauge, with which to measure the external diameter of the case. Write upon it, "⅞ space."

Procure some 60 lb., 70 lb., or 84 lb. imperial brown paper: the size of a sheet will be 29 inches by 22½. Cut a sheet into 4 equal parts, each 14½ by 11¼: paste the 4 pieces on one side, and lay them one on another, with the pasted face upwards, putting the fourth piece with the pasted side downwards, upon the pasted side of the third piece. Turn them over: take off the now top piece, and lay it flat on the near edge of a table, pasted side upwards. Take the former, fig. 7, and paste the tube all over. Lay it along the edge of the paper, bend the paper over with the fingers of both hands, and roll it tightly up, until the external diameter of the case about fits the gauge, fig. 8. If the paper should be too long, of course a piece must be cut off; if it should not be long enough, more must be added, taking care to bind in the second piece with 3 or 4 inches of the first piece; for if the whole of the first piece be rolled up before beginning the second, the latter, when dry, will probably slip off, and spoil the case. The case having been rolled up, take the handle of the former in the left hand, lay the case flat on the near side of the table, take the rolling-board, fig. 6, in the right hand, press the front part of it on the case, and drive it forwards 5 or 6 times, like a jack-plane, letting the handle of the former slip round in the left hand. This will tighten the case, and render it, when dry, as hard as a book-cover.

The former must always be pasted, before rolling a case, to prevent its sticking. It should, likewise, be wiped clean with a damp sponge, before being laid aside. Brass tubes keep clean a much longer time if lacquered. To lacquer them, clean them with very fine glass-paper; make them hot by the fire, till you can just bear them on the back of the hand; then, with a camel's-hair pencil, wash them over with thin lac solution. The cases may be either 14½ or 11¼ inches long; but 11¼ is the best, for when the cases are too long, the fuse, as it approaches the bottom, is apt, if slow, to smoke; if fierce, to set the top of the case in a flame. If the learner decides upon 11¼ inches, the former and rammer may each be 2 or 3 inches shorter.

After the first case has been rolled up to fit the gauge, it may be unrolled, and the paper measured. Future pieces of the same quire of paper can then be cut of the right size at once, so that the case will fit the gauge without further trouble.

A large slab of slate is convenient for rolling upon; but a smoothly planed board will answer every purpose.

When a number of cases are finished, hitch a piece of flax two or three times round each of them, and hang them up to dry, in a place free from draught, that they may not warp.

Flax is sold in balls; the thick yellow, at $2d.$, is the best. It is named, indifferently, flax, or hemp. It is much used by shoe-makers; and is sold at the grindery, or leather shops. Two or three thicknesses of this, waxed, or drawn through the hand with a little paste, is very convenient for passing round the necks of small choked cases, tying cases on wheels, &c.

To Make a Roman Candle Star.

Take the former, fig. 1, which, as said before, is $1\frac{1}{2}$ inch long; have a cylindrical piece of turned wood, box, beech, or mahogany, fig. 2, about 2 inches long, and of a diameter to just fit easily into fig. 1. At a point a, at the distance of about $\frac{7}{8}$ of an inch from the end d, with a bradawl, or very small gimlet, or nosebit, make a hole, and drive in a piece of brass wire, to project just so much as to prevent the tube slipping over it. A piece of a brass rivet, such as used by shoe-makers, is convenient for the purpose. The part with the head on is best; a quarter of an inch length will be sufficient, filed or cut off with the nippers. It is evident that upon inserting fig. 2 into the tube fig. 1, a vacant space of $\frac{5}{8}$ of an inch will be left at the bottom. Fig. 3 is a piece of turned wood, or, better still, of turned brass, exactly like fig. 2 without the side-pin a. Now to pump a star, insert fig. 2 in fig. 1; press the tube into damped composition, turn it round, and withdraw it. Rest the tube on a flat surface, insert fig. 3, and give it two or three taps with a small mallet, like fig. 26. A convenient size for the mallet is $1\frac{1}{2}$ inch square, 3 inches long, with a turned handle. The mallet is best made of beech or mahogany. The slight malleting consolidates the star, and prevents it from getting broken in charging; it will compress it to about $4\frac{1}{2}/8$ or $\frac{9}{16}$ of an inch in height. Push it out and set it by to dry.

Stars are best made in summer, and dried in the sunshine; when dry they should be put into clean pickle-bottles, furnished with tight-fitting bungs. A piece of wash-leather passed over the bottom of the bung, gathered up round the sides, and tied at the top like a choke, makes a good stopper. Shot, shaken up in bottles, with water, soon cleans them.

To Damp Stars.

Stars containing nitrate of strontian must be damped, either with lac solution, or wax solution; anything containing water destroys the colour. Nitre stars may be damped with gum water, dextrine solution, or thin starch. Most other stars with either of the solutions. Crimsons and greens *will* mix with boiled linseed oil, but they cannot then be matched, as oil renders meal-powder almost uninflammable. With all stars, not a drop more of the solution should be used than is sufficient to make the composition bind; and it is advisable not to damp more than half an ounce at a time; this is particularly the case in using the lac solution, as it dries rapidly; and if a large quantity of composition is damped, and gets dry, and has to be damped over and over again, it becomes clogged with the shellac, and the colour is deteriorated. If it should get dry, and require a second damping, it is best to use pure spirit only, the second time.

Before mixing compositions, every article should be as fine as wheaten flour, and perfectly dry. Nitrate of strontian, if purchased in the lump, should be set over the fire, in a pipkin; it will soon begin to boil in its water of crystallization; it must be kept stirred with a piece of wood, till the water is evaporated, and a fine dry powder left. A pound of crystals will yield about 11 ounces of dry powder, which should be immediately bottled. Even then, if used in damp weather, it is best dried again, and mixed with the other ingredients while warm. This second drying may be in a 6-inch circular frying-pan.

Articles, separately, may be reduced to powder, with the pestle, in a mortar. See that it is wiped clean every time, as there is danger of ignition with chlorates and sulphurets. When the articles are to be mixed, they may be put into the mortar, and stirred together with a small sash-tool. A $\frac{3}{8}$ inch is a convenient size. The mixture must then be put into a sieve, and shaken in the usual way; or it may be brushed through with the sash-tool. Return it to the sieve, and brush or shake through again. As it lies in a heap, level or smooth it with the blade of a table knife, or any straight-edge; if thoroughly mixed it will present a uniform colour; if it appears darker in one part than in another, it must be sifted again. A sieve with a top and receiver is very desirable, as nearly all mixtures are either black or poisonous;

the dust from star mixtures is very injurious to the lungs. If a top and receiver cannot be readily purchased, both may easily be constructed out of a sheet of millboard, fastened with a bradawl and waxed yellow flax, and neatly covered with paper.

Mixtures may be damped on a Dutch-tile, a marble slab, or a slate without a frame. They may be stirred about with a dessert knife, pressed flat, and chopped, or minced as it were, and again pressed flat.

To Make Lac Solution.

Put ½ an ounce of flake shellac into a tin pot, and pour upon it a quarter of a pint, or 5 ounces of methylated spirit; or, preferably, a like quantity of wood naphtha. Let it stand for about a day, stirring it occasionally till dissolved. Then half fill a basin with boiling water; set the tin containing the lac, in it, and leave it till it boils and curdles. If the water does not remain hot long enough to make it boil, set it in a second basin of boiling water. As soon as it has curdled, remove it; and when cold, pour it into a vial, and cork it. Spirit must never be boiled over a fire, nor near one, as the vapour might inflame. Keep the pot, therefore, while in the hot water, at a distance from a fire, or flame of a lamp or candle.

To Make Wax Solution.

Put into a vial ½ an ounce of white wax, (bleached bees' wax), pour upon it 5 ounces of mineral naphtha, (coal or gas tar naphtha), keep it tightly corked.

To Make Stearine Solution.

Dissolve a piece of composite candle in mineral naphtha, in the same way. Mineral naphtha must not be used near a candle or fire, as it gives off an inflammable vapour, at less than 100° Fahrenheit.

To Make Gum Solution.

There is no better way of preparing this than simply to put cold water upon gum arabic, and let it stand till dissolved. If for sticking purposes, as much water as will just cover the gum will be sufficient; but, for making quickmatch, 1 ounce or 1¼ ounce of gum to a pint of water. If required in a hurry, put the gum into *cold* water, in a pipkin, or tin saucepan, set it on the fire, make it boil, and keep stirring till dissolved. When cold, bottle, and cork it.

To Make Dextrine Solution.

Take half an ounce of dextrine, and 5 ounces, or a quarter of a pint of cold water, put the dextrine into a cup or basin, add a little of the water, and mix it well with a teaspoon, rubbing it till all is dissolved; then add the remainder of the water, stir well together a second time, pour it into a vial, and cork for use. Dextrine, wetted to the consistency of honey, may be used instead of thick gum-arabic water, for pasting. For this purpose it is advisable to keep either in a wide-mouthed bottle, and to set the bottle in a gallipot containing a little water; the brush, a camel's-hair pencil, or very small sash-tool with ⅓ of the bristles cut away on each side, to render it flat, can then be kept in the water, when not in use; this will prevent it, on the one hand, from becoming dry and hard; and, on the other, from getting clogged and swollen. It can be squeezed between the thumb and fingers, when wanted for use. The flat gum brushes now sold, bound with tin, are not pleasant to use, as the tin oxidises, and turns of a disagreeable brown colour. If there is a difficulty in obtaining a graduated water measure, one sufficiently correct for pyrotechnic purposes may be made with a vial. Paste a narrow strip of paper up the outside of the vial, weigh 4 ounces of water in a cup, in the scales: pour it into the vial, mark the height, and divide it into 4 equal parts, for ounces; of course, it can be graduated into half and quarter ounces, and increased, if large enough, to 5 or more ounces. A gallon of distilled water weighs exactly 10 pounds. Consequently, a pint of pure water weighs a pound and a quarter. This is also near enough for spirit, though, of course, spirit is a trifle lighter. Doctors' vials are often marked with ounce divisions.

To Make Paste.

Paste is most economically made in a zinc pot, which may be 4 inches deep, and 3½ inches diameter. Any zinc worker will make one to order for about 6*d.* Put into it 2 ounces of wheaten flour, add a little cold water, rub the two together with a spoon till smooth and free from lumps; pour in more water till the pot is full within about an inch; set the pot in half a saucepanful of water, put it on the fire; make the water boil, and keep it and the paste boiling for 4 or 5 minutes, stirring the paste the while. Remove it from the fire, and set it by to cool. The paste is to remain in the zinc pot, in which it will keep good for a length of time, and beautifully white.

Some recommend alum in paste, I think it best avoided, especially in cases intended to receive coloured fires. Alum is a double salt, a sulphate of alumina and potassa; it has an acid reaction; and, coming in contact with chlorate of potash and sulphur, may cause spontaneous combustion. A drop of sulphuric acid instantly ignites stars containing them. At theatres, the clown sometimes fires a cannon, with what appears to be a red hot poker; but which, in reality, is only a piece of wood, painted red. A mixture is made of chlorate of potash and sulphur, or sugar, a glass bead is filled with sulphuric acid, and the hole stopped up with wax. This is laid in the mixture, and when it is struck with the poker, the liquid escapes, and inflames the potash and sulphur. Sulphate of copper is a particularly dangerous salt, and must never be used, as it is almost certain to cause spontaneous combustion. Chertier, to whom pyrotechny otherwise owes so much, introduced an empirical preparation, by dissolving sulphate of copper in water, together with chlorate of potash, drying it, and wetting it with ammonia: but this, however dried, when again wetted, turns litmus paper red. Practicus has named it Chertier's copper. I discommend its use.

Two paste brushes will be sufficient for an amateur, sash-tools, one about an inch diameter, the other smaller for light purposes. Let them stand in the paste. If they get dry, the bristles fall out. For convenience, one may be kept in the paste, and one in water.

Dry clay, powdered and sifted as fine as possible, is used for plugging, or stopping up the bottoms of cases. I have, for some time, discontinued its use, and employ plaster of paris in preference.

Directions will be given for each, so that the learner can adopt which he pleases: but plaster is infinitely preferable. It is an American improvement.

Roman Candle Scoops.

No species of fireworks require greater care in their construction than roman candles. In the first place the stars must be fierce, that they may light thoroughly: next, they must not be driven out with too great velocity. For this purpose the blowing-powder must be carefully adjusted. The stars, also, must be of so easy a fit that when put into the case they may fall to the proper depth of their own accord. If they require pushing, they are too tight, and will probably be blown out blind. When made as directed they will, necessarily, be of an easy fit, as they will be of the inner diameter of the brass tube, while the bore of the case is equal to its external diameter.

To regulate the blowing powder, prepare a number of little scoops, like fig. 5, which is about the right size for the bottom star. They are formed of pieces of tin, zinc, or copper. Cut a long strip of tin, $\frac{1}{2}$ an inch broad; cut this across into 7 pieces, of the following lengths, $1\frac{5}{8}$, $1\frac{6}{8}$, $1\frac{7}{8}$, 2, $2\frac{1}{8}$, $2\frac{3}{8}$ and 4 inches. Round off the corners. Take a piece of brass wire, or stair-rod, about $\frac{1}{4}$ inch diameter, and with the wooden mallet, before mentioned, fig. 26, bend each of the pieces round the rod into a half cylinder, or gutter. Take up the smallest, and hold $\frac{5}{8}$ of an inch of the end of the stair-rod in the end of the semicylinder, to keep it open; put the other part from a to b, fig. 5, in a vice, and pinch it up; it will assume the form represented; the bowl part will be $\frac{5}{8}$ long, and the handle 1 inch long. Make the bowl of the next scoop $\frac{6}{8}$ of an inch long, the next $\frac{7}{8}$, and so on; the handle will always be 1 inch long. The last, for the top star, will have a bowl of 3 inches. The smallest scoop ought to hold as much grain powder as will weigh about $\frac{1}{12}$ of the star; but to have the scoops accurate, it will be necessary to charge a roman candle, fire it, and observe whether the stars go a uniform height. For measuring the interval fuse, or fuse between the top of one star and the bottom of the next, a large scoop of the size of fig. 10 will be required. The tin may be an inch broad, and the bowl part $2\frac{1}{2}$ inches long, bent round the rammer, fig. 4. To adjust it, take a roman candle case, fit it on the foot, fig. 9, which is a piece of wood, or brass, turned with a tenon to fit tight in the bottom of the case. Fill the scoop, and strike it level, with a straight-edge; empty it into the case, rest the foot on a flat surface; insert the rammer, fig. 4, and jolt it up and down, a dozen times, or more, lifting it about ½ an inch at a time; put in another scoopful, and jolt it in like manner. If the 2 scoopfuls, thus compressed, fill an inch of the case, the scoop will be correct. If more or less, the scoop must be shortened, or lengthened, accordingly.

A piece of writing-paper may be pasted and wound twice round the handle of each scoop, as from a to b, fig. 4. One dot can be put upon the scoop, for the first, or bottom, star; two dots, for the second scoop, &c., or any memorandum can be written upon them, for future guidance. Should they get soiled, they may be cleaned with a soaped damp piece of sponge.

Gunpowder, for fireworks, is used in two forms; meal-powder, and grain-powder. Meal-powder is a fine black dust, and is employed in all cases of mixing. Grain-powder is of three kinds, F, FF, and FFF, fine, double fine, and treble fine. FFF is best for crackers, simply because it runs rapidly down the pipes: for driving stars, shells, &c., F will be sufficient; but FFF may be employed: FF need not be purchased. All kinds of powder may be obtained of Pinnell, 214, Whitechapel Road. If, in any place, there should be a difficulty in obtaining meal-powder, F grain-powder may be crushed in a leather bag, by laying the bag on a hard surface, and beating it with a hammer. The leather should be of the same kind as shoes are made of.

To Charge Roman Candle Cases.

Pour some F grain-powder into a wooden bowl, or platter, represented by fig. 11. Round the edge lay the little blowing-powder scoops, side by side, beginning with the smallest at a, the next at b, and so on to g. Put some roman candle fuse into a large tin scoop, made to stand on a flat bottom, like fig. 12, the same, in shape, as used by tea-dealers; and, on the right-hand of it, lay the charging-fuse scoop, fig. 10. If the roman candle is to contain different-coloured stars, set seven in a row, in the order desired. When the cases are intended to be fired in threes or fours, the stars in one may be all blue, in another crimson, in another green, in another white. Fit the foot, fig. 9, in the bottom of the case, put in a scoopful of clay, insert the rammer, fig. 4, and jolt it till the clay is well compressed. The clay should fill ½ an inch. This being done, invert it, and shake out any little dust that may remain. Put in the little scoopful a, of F grain-powder; then lay the scoop at A. Now put in a star. As previously stated it ought to fall of its own accord; but make sure that it has reached the blowing-powder, by putting in the rammer. Having ascertained this, put in a scoop of fuse, fig. 10; lay the scoop on the left of fig. 12; insert the rammer and jolt it; put in another scoop of fuse, fig. 10; lay the scoop on the right of fig. 12; insert the rammer and jolt it, as before. Then proceed with the scoop b of grain-powder, and lay it at B, and so on, till the case is filled. The fuse on the top star is best driven in with a short solid rammer and mallet, as it is difficult to jolt the long rammer in so small a space. The last ⅛ of an inch, near the mouth of the case, should be fine meal-powder, as it binds better than the roman-candle fuse, and also blows off the leader pipe.

The blowing-powder scoops, having been laid at A, B, &c., all that is required is to turn the bowl or platter, a little round to the left, and they will come in rotation, ready for the next case. Also, by putting the scoop, fig. 10, alternately to the left and right of the scoop, fig. 12, it will always be known whether the proper quantity of fuse has been put in.

Coloured stars, from their fierceness, have a tendency to burn in the cases. This defect may be remedied by putting upon each star a small scoopful of Starting Fire, No. 1, before putting in the interval fuse; as much as will fill round the sides of the star. This composition is somewhat fiercer than would suit for the regular fuse, so catches the blowing-powder sooner.

A roman candle is well charged when the stars isochronise, or come out at equal intervals of time: they should, also, theoretically, ascend to equal heights; but, with coloured stars, this cannot be perfectly insured, as some shrink more than others in drying, and, of course, fit more loosely; some are heavier, some fiercer than others.

The interval fuse must always be driven in at twice, never at once. Each star, with its blowing-powder and fuse, occupies about an inch and a half; perhaps a trifle more.

Instead of driving in clay at the bottom, plaster of paris may be used, and then the foot, fig. 9, will not be required. Have some plaster of paris in a wide-mouthed bottle; a glass of cold water with a salt-spoon in it; and a number of pieces of paper, about 4 inches square. Put a small quantity of the plaster on one of the pieces of paper; indent the middle with the finger; put to it a little water, and work it up with a dessert-knife. Just as it gets to the consistency of mortar, and is about to set, mould it with the fingers, to the shape of a cork; push it into the end of the case; rest the case on a flat surface; insert the rammer, and give it two or three slight jolts; turn it round a few times, and withdraw it. If the plaster sticks to the end of the rammer, it shows, either that you have used the plaster too wet, or have not turned the rammer round a sufficient number of times.

No more plaster must be mixed at a time than will suffice for one case. When plaster has once set, it cannot be mixed up a second time; therefore take a fresh piece of paper, and let the knife be cleaned every time. It is advisable to have two dessert-knives, then one can be used with which to scrape the other. As much plaster should be used as will fill the case up about $\frac{1}{2}$ an inch. They must be set by to dry; their not requiring the use of the foot will be found a great convenience.

Roman candles are usually made from $\frac{3}{8}$ to $\frac{6}{8}$; but $\frac{5}{8}$ is a very satisfactory size. If a roman candle is intended to be fired singly, twist a piece of touchpaper round the mouth. If the cases are intended to be fired in threes, fours, &c., to form a bouquet, or to be placed round a mine, jack-in-the-box, or devil-among-the-tailors, omit the touchpaper, and envelope the case in double crown, made to project an inch beyond the mouth, to receive the leader, or quickmatch.

A steel-pen inserted, nib backwards, in the end of a small paper tube, rolled round the end of a penholder, makes a neat little scoop. It may be fastened in with a little plaster of paris. A scoop may also be made with a quill.

Cracks in wooden bowls may be stopped with the same material; and, if painted over with linseed oil, after getting dry, will remain waterproof for a long time. A screw may be made to hold in brickwork, by drilling a hole in the brick, and pushing in the screw, covered with plaster.

To Make Touchpaper.

Dissolve ½ an ounce of nitre in ½ a pint of hot water. Procure some 12 lb. double crown blue; cut each sheet into four equal parts, 15 by 10. Lay them smooth upon each other, and, with a sash-tool dipped into the nitre solution, wash them over on one side, and hang them up to dry.

To Make Slowmatch.

Dissolve 1 dram of nitrate of lead in ½ an ounce of boiling water. Cut a sheet of blotting paper into six equal parts, and wet them on both sides, with a sash-tool, with the solution. When dry, paste a piece all over, and upon it smoothly press another piece; upon this, pasted, put a third piece; and so on, till all the six form a stiff board. Lay them under a heavy weight; and, when dry, with a sharp knife and straight-edge, cut the whole into strips ¼ of an inch broad. Four inches will burn about a quarter of an hour. Narrow tape, boiled in the solution, makes excellent slowmatch.

To Make Quickmatch.

Put into a pan 1 lb. of grain-powder, or meal-powder; pour upon it some thin hot starch, and stir it well about, breaking all lumps, till the mixture is of the consistency of paint. Procure some lamp-cotton, such as forms the wicks of candles. It will probably consist of sixteen or twenty-four strands. Divide it carefully into lengths of eight strands. This is not so easy a task as might appear. The best way is to act the reverse of a man spinning string. Divide the end of the cotton, say of sixteen strands, into two of eight each; fasten them to two screw-hooks, a few inches apart. Take one in each hand, and walk backwards, gently pulling them apart, and when they catch, untwist them: with care they will separate without breaking or entangling. Drop the end of one of the pieces into the pan; and, as it keeps falling, coil it round and round in the mixture, and press it down with an iron spoon, until as much is pressed in as the quantity can saturate. Be very particular that it is thoroughly soaked. Have ready a wooden frame, fig. 13, of deal, resembling a swing looking-glass, with the glass taken out. It may be 5 feet by 4. The frame is to be supported on pivots between two uprights. Fix a nail or hook at the left-hand corner of the frame, and tie the end of the cotton to it which has been hanging outside the pan. Get a person to slowly turn the frame and hold it steady. Take hold of the cotton in the right hand, shut the hand, and allow the cotton to slip gradually and slowly through it, as the frame is turned; squeeze it very gently, so as to allow it to come out well coated, and contrive to make it as round as possible. When all the cotton is wound upon the frame, spread some sheets of paper—old newspapers—on the floor; at each corner place a brick; lay the four corners of the frame upon the four bricks; sift dry meal-powder all over the match, turn it over and sift over the other side. Prop the frame against the wall, and leave it to dry. One ounce of white starch will be sufficient for a pint and a half of water. Rub the starch up smooth with a little of the water, then add the rest, and boil it.

A New Method.

Put into a gallipot, or basin, some hot starch, made as before directed, or some cold gum-water, or cold dextrine solution; and with a small stencil-brush, or a $\frac{3}{8}$ sash-tool, cut across the middle to make the bristles short and stubby, stir in some meal-powder, till it is well mixed and looks like black paint. To avoid repetition, it will be convenient to refer to this under the name of Meal-paste. Take two towel-horses, fig. 14, and set them parallel 5, 10, or 20 feet apart. In the top rail of each, drive four nails $\frac{1}{2}$ an inch asunder. Fix the dry cotton to the nail a, carry it across to the opposite nail b, pass it round the second nail c, bring it to the opposite second nail, and so on, till the cotton lies in four parallel lines, like the strings of a harp placed horizontally. Hitch the cotton, without cutting it, to the hook of the weight, fig. 15. This weight is made by taking a piece of brass tube, 1 inch diameter, and 4 inches long. Stop one end with a bung, fill it with melted lead; and before it sets push in a lucifer match, having previously cut off the priming. As soon as the lead is cold, pull out the match, which having been partially burnt smaller, will come out easily, and in the hole left by it screw a cup-hook, as drawn. Invert the weight, pick out the bung, and fill its place with more melted lead. Now procure two pieces of planed deal board; one 4 inches square, the other 6 inches square. With the left hand hold the smaller piece close underneath the cotton; and with the sash-tool or stencil-brush, work the meal-paste *well into the fibres of the threads*, pressing the cotton on the board till it is thoroughly soaked, and rolling it over, laterally, to make it as round and smooth as possible. It is best to begin on the left, and work towards the right. If the wetting slackens the threads, pull them tight. Now brush some meal-powder through a fine sieve, to free it from lumps; put a tablespoonful or two upon the larger board; hold it close under the four threads, as the other, with the left hand, and move it laterally forwards and backwards, and down the whole length, at the same time brushing the meal over the threads, with a soft, dry, sash-tool, till they are smothered, and giving them an occasional jar to shake off the superfluous meal. By a little practice they may be made as smooth and as round as a piece of wire. Leave the weight hanging to them till they are dry. Instead of four nails on each rail, a dozen may be put; and if the towel-horses are set 12 yards apart, 144 yards may be soon made. In this case, three or four heavier weights would be required. These contrivances, however, are by no means necessary for an amateur; four hooks or nails opposite four others, anyhow supported, will be sufficient; and six or eight feet apart is a good distance. A small quantity may thus be made one day, and a small quantity another; and for this purpose it is best to use gum-water, as it is always ready, and a little can be added to the dry left the day before, and a little fresh meal stirred in. Three or four threads of white darning-cotton, which is of two twists, make very good match; knitting or crochet cotton, which is of three twists, produces, alone, excellent match; two or three pieces of the first, or two of the latter, put into a leader pipe, side by side, blow through with a violent report. Match may also be made of the very narrowest white tape, $3/16$ of an inch broad; this, from its flatness, is peculiarly suited for enveloped stars. For general purposes, however, lamp cotton is decidedly the best, as it is most loosely twisted, and therefore the most absorbent. If match is liable to be exposed to damp, it should be cased in thick leaders, and be prepared with starch; in other respects, nothing can surpass gum-water. Of course iron pound-weights, with a ring in them, will answer the purpose, or an iron pestle tied to the cotton, or anything heavy; but the kind I have advised are most convenient. They will weigh about a pound and a quarter each. The weight may hang over a chair-back.

Match, to be perfect, ought, when cut across, to look black throughout; it should, also, be stiff, straight, and round; but, to test it, cut off about 10 inches; put one end of this into a leader pipe, so that 5 inches will be in the pipe, and 5 out. Hold the end of the leader with a pair of tongs, or lay it on the ground; light the naked end. If the match is good it will burn gradually, though swiftly, till it gets to the pipe; it will then blow through with a bang. The nearer the leader pipe fits the match the better, only it ought to be large enough to allow it to go easily in, without force: if laid in a roman candle case it would hardly puff: the smaller and smaller the tube becomes the louder and louder the report: the increase of power being in the inverse ratio of the diminution of space. A train of gunpowder laid in the open air, or confined in a tube, comports itself in the same way, as is well known. Match, when dry, should be kept straight; for, if it gets broken, it acts like a cracker, snapping at every break.

A piece of zinc rainwater pipe, of suitable length, furnished with a bottom and lid, or a couple of bungs, is convenient for keeping it in; but a far better contrivance is a deal box, 5 feet long, 3 inches deep, and 3 broad, made of ½-inch pine. The lid of this box can be readily furnished with three or five hinges, like a piano, only made with string instead of brass. Take two pieces of string, fig. 57, and tie them in a knot near the end, as at k. At x, fig. 58, make a bradawl hole through the back of the box, near the top, and push the strings through it; make another hole through the lid, put the string through it, and tie in a knot on the top, as at z: and so with the other four. This will render the explanation easier to understand; but, practically, a single piece of thicker string is best, if you can manage to tie the second knot in the exact spot you wish. The knots must not be too close to each other, as the lid requires a little play. For safety, the box ought to be furnished with a lock.

ROCKETS.

Rockets are charged in choked cases, on a spindle, to leave a hollow up the middle, through which the fire may be communicated to nearly the whole of the composition at once: this causes the sudden generation of an enormous quantity of elastic vapour, which, being unable to escape instantly through the contracted aperture left for its exit, exerts its pressure in the contrary direction, and hurries the rocket forward. A stick, attached to it, guides it, like the rudder of a vessel, or tail of a bird, or fish; while its weight and leverage keep the centre of gravity a little below the case, and prevent the rocket from pitching over. Its manufacture, therefore, from the commencement of cutting the paper for the case, to its finish of fitting on the stick, requires an accurate adjustment of all its parts. Disregarding the trade names of pound, ounce, &c., which, now that moulds are dispensed with, are useful only for enabling the makers, from tradition, to understand each other; the internal diameter, instead of the external, as formerly, may be selected, from which to compute the relative measures. Taking the bore of the case as unity, the proportions will be as follow:—

1	Internal diameter of case.
$1\frac{1}{2}$	External diameter.
8	Length of case.
6	Length of spindle.
$\frac{3}{5}$	Bottom diameter of spindle.
64	Length of stick.
$\frac{1}{2}$ by $\frac{1}{2}$	Thickness and breadth of stick.

These proportions are most readily calculated by taking the diameter in eighths of an inch. Selecting, for instance, a $\frac{6}{8}$ rocket, we have, $\frac{6}{8}$, inner diameter; half as much again, $\frac{9}{8}$, outer diameter.

Taking the numerator as inches: 6 inches, length of case; $\frac{3}{4}$ of this, or, which is the same, $\frac{1}{2}$ the outer diameter, as inches, $4\frac{1}{2}$ inches, length of spindle; putting 20 for the denominator, instead of 8 (8 being $\frac{2}{5}$ of 20), $\frac{6}{20}$ of an inch, bottom diameter of spindle; 6 × 8, (numerator multiplied by denominator) = 48 inches, length of stick; $\frac{3}{8}$ by $\frac{3}{8}$, size of stick.

The following table exhibits the usual sizes in inches.

Name of Rocket.	Inner Diameter.	Outer Diameter.	Length of Case.	Length of Spindle.	Bottom Diam. of Spindle.	Length of Stick.	Size of Stick.
10/8	10/8	15/8	10	7½	10/20	10 times 8=80	5/8 by 5/8
9/8	9/8	13½/8	9	6¾	9/20	9 times 8=72	4½/8 by 4½/8
8/8	8/8	12/8	8	6	8/20	8 times 8=64	4/8 by 4/8
7/8	7/8	10½/8	7	5¼	7/20	7 times 8=56	3½/8 by 3½/8
6/8	6/8	9/8	6	4½	6/20	6 times 8=48	3/8 by 3/8
5/8	5/8	7½/8	5	3¾	5/20	5 times 8=40	2½/8 by 2½/8
4½/8	4½/8	6¾/8	4½	3⅜	4½/20	4½ times 8=36	2¼/8 by 2¼/8
4/8	4/8	6/8	4	3	4/20	4 times 8=32	2/8 by 2/8
3/8	3/8	4½/8	3	2¼	3/20	3 times 8=24	1½/8 by 1½/8

To Make a 6/8 Rocket.

Have the former, fig. 7, a brass tube 6/8 of an inch external diameter; the gauge, fig. 8, with a 9/8 aperture; procure some imperial brown paper, 70 lb. or 84 lb., the thicker and heavier the better. The best kind is made of old ropes, is air-dried, and rough. This is not easily obtained now. The smooth machine-made, cylinder-dried answers very well. The sort used for laying under carpets, and which is 5 feet broad, is almost equal to the original rough imperial, and should be procured, if possible, in preference to the smooth. Cut it into strips 6 inches wide. Paste the pieces well, and roll the cases as hard as possible, with the rolling board fig. 6, till they fit the gauge. Lay them by for a few hours to get partially dry. They must, then, be choked or strangled, about half a diameter from one end of the case, that end that lay nearest to the left hand in rolling, till they assume the shape of the neck of a vial. For this purpose have a cylindrical piece of deal, alder, or any kind of wood, fig. 24, about 9 or 10 inches long, made to fit easily into the case; cut it into 2 pieces, a and b; b may be an inch and a half long; round off the ends, just cut. In a, fasten a screw, the solid wire, or uncut part of which, is ¼ of an inch thick; saw, or file off the head, and slightly taper the part projecting; in the piece b bore a hole to just fit the wire of a. Slip b upon a, and push them into the case, so that the interval between the two reaches within about ½ an inch of one end of the case; slightly draw out b, to leave a neck, or hollow, round which to form the choke; fix a staple, or screw-eye in a post: tie to it one end of a piece of cord, about a yard long, and ⅒ of an inch thick; fasten the other end round the middle of a stick, for a handle; take hold of the stick with the right hand; hold the case in the left; pass the cord round the space left vacant and gradually tighten it, turning the case round and round, so as to pinch it in equally on all sides, till it assumes the shape of fig. 23. Hitch a piece of string, about ¹⁄₁₆ of an inch thick, a few times round the neck, until it is filled up flush with the other part of the case, remove the choker, and hang the case up to dry. A hitch is made by simply bending the string backwards into a loop, like fig. 22; passing it round the choke; pulling it tight; looping it again, again passing it round the choke, and so on. The first loop is called a half hitch, and will not hold of itself; the second loop completes the hitch, and makes it hold: the remaining hitches are for filling up the choke, and restoring the cylindrical shape. Practise, by hitching the string upon your thumb, 4 or 5 times: it will hold after the second loop; but push it off at the end, and it will all fall loose again.

A case must not be choked when wet, or it will tear; nor must it be too dry, or it will be difficult to choke it at all. The drier, however, it can be choked, the better. Experience is the only guide. If properly dry, the wrinkles of the choke will be small, and perfectly regular. Should the string stick to the case, chalk it; but this is not likely to occur unless the paper is too wet. Have a foot, fig. 16, turned of ash, or beech, or box, with a hemispherical nipple, ⅝ of an inch diameter, as drawn. Bore it with a twist drill, or nosebit, to the depth of an inch and a half. Procure a brass, iron, or steel wire, but preferably brass, 6 inches long, and ³⁄₁₀ of an inch diameter, and perfectly straight. Cut, on one end of it, a screw, 1½ inch long, fix it in a vice, wet the screw with glue, and screw the foot on. File the 4½ inches gradually tapering. The object of tapering it is simply to make it deliver. A cylindrical hollow up the rocket would answer as well, but the spindle could not be got out. The more conically true it is tapered, of course, the better. Finish it off with a very fine file, and smooth it with glass paper. The block into which it is screwed may be larger than drawn; the bottom should be turned slightly concave, to make it stand firm, on the same principle as the bottoms of plates, cups, &c., are made with a rim. The block and spindle are better if cast in gun-metal, in one solid piece: the pattern to give to the caster should be in one solid piece of wood. After casting, the rough parts must be filed smooth; not many turners will be found willing to undertake to turn it.

The next articles required are a setting-down piece, fig. 17; three hollow rammers, or drifts, figs. 18, 19, and 20; and one solid rammer, fig. 21. They are simply cylindrical pieces of wood, turned with a head, to bear the blows of the mallet. Beech, or box free from knots, will answer. The lengths of figs. 18, 19, 20, and 21, are to be 8, 6, 4, and 2 diameters, respectively, exclusive of the head; that is, 6, 4½, 3, and 1½ inches. The hollow drifts are to have a cylindrical, not conical, hole, bored up them, with a nosebit, or twist drill, to within one inch of the handle, so as to clear the spindle by ½ an inch, to allow of any dust being driven up it. The hollow in figs. 17 and 18 must be large enough to fit the spindle loosely at a; in fig. 19, a trifle smaller, to fit loosely at b; and in fig. 20, a trifle smaller still, to fit loosely at c.

The next requisite is a mallet, fig. 25, which may be of ash, or beech. It may be a cylinder, 5 inches long, and 3 inches diameter, with a handle about 5 inches long, and 1 inch diameter. Let it be turned at the end slightly concave, like the bottom of fig. 16, that it may be set to stand upright, like a wine bottle. Or the head may be made 3 inches square, like fig. 26; or, a small carpenter's mallet, about a pound and a quarter weight, will answer.

To Charge Rocket Cases.

The first requisite is a solid block of wood, 6 or 8 inches square, and 18 or 20 high. This is indispensable. A piece of an old oak gate post, answers well. It must be set upon the ground, or on a flag stone, not on a floor. It is impossible to ram a rocket properly on the floor, because of the vibration. It is, also, necessary to sit, not stand, at the work.

Put the setting down piece, fig. 17, into the case; press the case over the spindle, and give the head a few blows with the mallet; this will smooth out the wrinkles of the choke, which is all that fig. 17 is used for. Now put in a very little powdered clay, and mallet it with fig. 18; as much clay as will reach up $\frac{1}{12}$ of an inch will be sufficient; its object is to preserve the choke from burning, and getting enlarged. One cannot be too precautious with rockets. Now put in a scoop of rocket fuse, insert drift, fig. 18, and mallet the fuse in firm, with about a dozen and a half blows, or till it offers a resistance to the hand. The blows must be light and numerous, not slow and heavy, like driving a post into the ground. 18 blows with a momentum of 3, will consolidate the fuse: 3 violent blows, with a momentum of 18, would perhaps bend the case, or drive the dust up into your eyes. The mallet need not be lifted above 4 or 5 inches at a time. If the rocket is not rammed firm throughout, it will, upon lighting, explode.

As soon as the case is charged about $1\frac{1}{2}$ inch, make a pencil or ink mark, round the drift, where it stands level with the top of the case, for future guidance; then charge another $1\frac{1}{2}$ inch with the second rammer, fig. 19, and mark it in like manner; proceed in the same way with fig. 20. It is obvious that if fig. 19 were used too soon it would get split by the spindle being driven up it, and the spindle would be bent or broken, hence the advisability of marking the drifts to know when to lay aside one, and take the other. Just before you get to the top of the spindle, put in the solid rammer to feel how high the spindle reaches near the top of the case; hold it by the thumb and finger to keep the distance, and mark it down the outside of the case, by indenting the case with the edge of the drift. Exactly $1\frac{1}{3}$ diameter, that is, in this case, exactly 1 inch, above this indentation make another mark: then as soon as you have covered the spindle, till you can no longer see it, with the use of fig. 20, begin charging with the solid drift, fig. 21, till the composition inside is level with the top mark. This being done drive in a little dry clay, till the case is full. Remove the rocket from the spindle, by giving it a turn or two round to the right, not backwards; then bore a hole through the clay, till you can see the composition, with a $\frac{3}{16}$ inch shell-bit. The shell-bit should be fixed in a handle, and kept for the purpose. It is not advisable to use a stock and bit, unless the bit is shielded, as it is apt to bore too deeply. The bit may be fixed in a handle, by boring a large hole in the handle, and pouring in melted lead, or pressing in plaster of paris.

Instead of driving in dry clay on the top of the composition, a little plaster of paris may be pressed in; this, when dry, will allow of a perfectly clean hole being bored through it; whereas the clay is apt to crumble, and chip out. The object of the clay, or plaster, is to prevent the composition, which, containing much charcoal, does not bind well, from getting disturbed, and the solid part diminished, which would cause the stars to be ignited while the rocket was ascending, or the fuse, perhaps, to blow through at the beginning. A piece or two of naked quickmatch is to be inserted in the hole through the clay or plaster, and a long piece is to be pushed up the core, or hollow, of the rocket, as far as it will go; it is, then, to be cut off flush with the mouth, and fastened to the side with a little dab of wetted meal powder, pressed on it with the blade of a knife. If the rocket is intended to be lit with a port-fire, take a circular piece of touch-paper, about 2 inches diameter, slightly paste it all over, lay it in the left hand, press the mouth of the rocket down upon it, and smooth the edges of the touch-paper up round the case. Otherwise, smear the end of the case with the sash-tool dipped into meal paste, and when dry, paste a bit of touch-paper round it, and twist to a point, like a squib. The appearance of the rocket is shown at fig. 36; the dotted line round the mouth shows the touch-paper.

In driving with the hollow rammers, it generally happens that a little of the fuse gets driven up the hole; this, if allowed to accumulate, is very troublesome to remove; it should, therefore, be knocked out every time, by holding the drift in the left hand, and giving the head a rap or two with the mallet.

The whole of the composition ought to be put in in about 12 scoops: try 2 or 3 scoops till you get one of the right size, then write upon the handle what-sized rocket it belongs to. These directions may appear minute, but they will save much trouble if attended to.

As it is convenient to know, beforehand, about what quantity of composition will be required for any particular rocket, the following formula will be useful:—

$E^3 / 9 =$ drams.

Where E denotes the size of the rocket, in eighths of an inch.

Required the quantity of fuse for a $6/8$ rocket.

$(6 \times 6 \times 6) / 9 = 4 \times 6 = 24$ drams = $1\frac{1}{2}$ ounce.

For a $3/8$ rocket?

$(3 \times 3 \times 3)/9 = 3$ drams.

The same weights denote the quantity of stars which the rocket will safely carry: thus an ounce and a half of stars may be put into the head of a $6/8$ rocket; and 3 drams into a $3/8$ rocket. Along with the stars is to be put in $1/9$ the weight of the stars of bursting powder; this may be pure meal powder; or a mixture of 8 meal powder, 1 fine charcoal, well sifted together: or half meal, half grain; thus the quantity of bursting powder for a $6/8$ rocket will be $24/9 = 2\frac{2}{3}$ drams; and for a $3/8$ rocket, $3/9 = 1/3$ of a dram. It is advisable to keep nearly to these directions, for the weight of the stars; but it is not necessary to be minutely exact; but to ascertain whether the head is too heavy or not, take a table knife, or any thing an inch broad, and lay the rocket flat and horizontally upon it in such a manner that the commencement of the head lies close to the back edge of the knife, the cutting edge lying towards the choke: if the head pitches over, it is too heavy, and some of the stars must be taken out; if it balances, it is correct.

Rockets should be fired from two staples, or two screweyes, fixed in a post, one near the top, the other half a yard below, as in fig. 70. They should never be propped against a wall, a chairback, a gate, or railing, as they might fall, especially on a windy night, after the touch-paper was lit, and before the fuse had caught. Every possible care should be taken in guiding them, as it is too late to think about any mischief they may cause, after they have once started.

In making rockets, it is essential, above all things, to have good nitre and charcoal. The best way, with fresh materials, is to weigh out as much nitre, charcoal, and sulphur, as will make one small rocket. Have the nitre as fine as possible, and dry it over the fire in a 6-inch frying pan, which should be kept for the purpose. If the rocket ascends well you will know that the articles are pure, and you can proceed to use them; but if the rocket does not rise, you may conclude the articles are adulterated, the nitre with salt, or that the charcoal is perhaps merely deal sawdust, burnt in a retort. In this case you must buy the nitre in crystals, and the charcoal in sticks. To powder the nitre, put it into a pipkin, pour on it a little water, set it on the fire, make the water boil, and keep stirring the nitre with a piece of wood, until it is dry and of a fine powder. Charcoal is best ground up in a coffee-mill. It must first be broken into small pieces, about the size of coffee-beans. After being ground it should be sifted through the sieve, the interstices of which are about $\frac{1}{40}$ of an inch square. To get a correct idea of this size, lay down a line on paper, an inch long: mark it off into eight equal parts by taking the divisions from a foot-rule; then carefully divide one of the eighths into five equal parts. But the best way is to make a square deal sieve, about 8 inches square, and 3 inches deep; then nail on the bottom a piece of perforated zinc, with quarter-inch flemish tacks. The perforations are circular, and should be 20 to the inch, measured diagonally; that is, a diamond-shaped inch, or inch rhombus, consisting of 20 oblique rows, each row containing 20 holes; $20 \times 20 = 400$. These holes will be about the same size as the ones mentioned in the sieve, because the solid parts between the perforations are also about $\frac{1}{40}$ of an inch. A strip of deal, $\frac{1}{4}$ of an inch thick, should be nailed round the bottom, to keep the zinc tight. A square box for a receiver, and another for a lid, should also be constructed, otherwise you will be smothered in sifting charcoal. Making use of the perforated zinc sieve just described, all the charcoal that goes through, fine and coarse, should be used for rockets. It need not be shaken much, but brushed through with a sash-tool. Construct a second sieve with 15 perforations to the lineal inch. Iron borings for gerbes should be sifted through this; use all that goes through. Construct, also, a third sieve with 9 perforations to the lineal inch. Nothing can equal iron and steel for making sparks; but neither will keep long from rusting, after coming in contact with nitre. Coke grains, about the size of pinheads, are a fair substitute. Beat coke into lumps about as large as peas, then grind it in a coffee-mill, and brush it through the 9-perforation sieve; sift the fine dust from what passes through: throw the dust away, and keep the grains. Porcelain may be powdered in an iron mortar, for gerbes; brush through the 9-perforation sieve; sift out the dust with a fine sieve; throw it away and keep the grains. The intense heat of the focus of the choke renders them incandescent; and, from their weight, they are projected to a considerable distance. They are inferior, however, to coke grains, as the latter are to steel filings. We may say, alliteratively; positive, porcelain; comparative, coke; superlative, steel.

Zinc may be obtained with half-inch perforations; a size useful for garden sieves, bottoms of soap boxes, &c.

Before grinding a fresh substance in a coffee-mill, it must be taken to pieces, brushed clean, and screwed up again. Fine lawn or hair sieves should be used for sifting chemicals; excellent sieves may be made with book-muslin: the cylinders may be 4 inches diameter, 3 deep; the muslin should be cut into a circular form, and hemmed round a piece of string; it may then be slipped over the drum or cylinder, and secured; or it may be pasted up and round the sides, and if above 4 inches diameter, two pieces of string may be crossed over the middle to strengthen it.

Charcoal may be made by putting some dry pieces of willow, alder, poplar, sycamore, maple, or almost any kind of wood, except the pine or turpentine tribe, into an old iron saucepan, covering them with perfectly dry sand, and setting the saucepan in the middle of a fire, to remain red hot till the wood is completely burnt through. Remove when judged sufficiently charred; and when cold, not before, pour away the sand.

Sulphur is used in the state of sublimed sulphur, sulphur sublimatum, or flowers of sulphur, and, when mixed with nitre, requires no preparation; but as it is always more or less impregnated with sulphuric acid, as is readily shown by testing it with litmus paper, it might, on coming into contact with chlorate of potash, cause spontaneous combustion. To prevent this, it is necessary to wash the sulphur. For this purpose put it into a pan, and pour upon it boiling water, in which some salts of tartar (carbonate of potash) have been dissolved; stir it well and break down all lumps. Let it stand to subside; pour off the supernatant liquor; fill up with cold water and let it stand, to again subside. Make a conical bag, fig. 33, with a piece of linen or calico, sewed at the top, round a ring or hoop of wire, or cane, or whalebone; fasten a string to it, by which to hang it up. Put the washed sulphur into it, and hang it under a water-tap; turn the water gently on, and let it drip all night; this will wash away every trace both of acid and alkali. Afterwards hang the bag up three or four days till the sulphur is dry; it may then be bottled, and kept exclusively for colours.

Oxalate of soda may be made thus—procure 3 lbs. of carbonate of soda, the common washing soda used by the laundress, not bicarbonate of soda; boil it up in a saucepan with just as little water as will suffice to dissolve it. Dissolve, in another vessel, 1 lb. of oxalic acid in boiling water, and pour it into a deep jar, capable of holding two or three quarts; a wash-hand jug will answer. Now put to this the dissolved carbonate of soda, with a table-spoon, a spoonful at a time. A violent effervescence takes place. The soda is to be slowly added till effervescence ceases. It should be tested with a strip of litmus paper, to see if the acid is perfectly neutralized.

To prepare litmus paper, dissolve ¼ of an ounce of litmus in an ounce and a half of water; when thoroughly dissolved, and the water is of a dark blue colour, take some white blotting paper, and with a sash-tool or camel's-hair pencil, go over it on both sides with the litmus solution. When dry, wet some of these prepared pieces, with the brush dipped into vinegar: this will turn them red. Dry, and preserve both. They may be cut into strips, half an inch broad; the blue strips will be tests for acids; the red, for alkalies. Wet a strip of the blue, and touch it with oxalic acid, it will turn red; wet a strip of the red, or the piece just reddened, with carbonate of soda, it will turn blue.

To make sulphuret of copper, procure some thin sheet copper, about as thick as a card; cut it into pieces, and put it into a crucible, with sulphur, a layer of sulphur, and a layer of copper alternately, till full. Set the crucible in a clear fire, and keep it red hot for an hour. Remove it; when cold, break it up, and grind it in a coffee-mill. Sift it in a lawn or book-muslin sieve as fine as possible. Half-a-pound of copper and a quarter of a pound of sulphur may be employed.

There is a black sulphide or sulphuret of copper produced by passing sulphuretted hydrogen through a solution of protoxide of copper: this is useless.

For want of a coffee-mill, charcoal may be beaten in a leather bag, with a hammer.

A variety of rocket fuses will be found in the Tables; the first is as good as any, and will answer for all sizes from $\frac{3}{8}$ to $\frac{12}{8}$. As a rule, the fiercer fuses, containing meal powder, may be used for small rockets; but are, by no means, necessary.

A rocket, when starting, makes a roar; but this is not on account of the fierceness of the fuse, but of the extent of the surface ignited. Rocket composition, laid in a train, burns very slowly.

Rocket Stars.

Rocket stars are made in three or four ways. First cut, or chopped, or naked stars. This mode is used for nitre stars only: chlorate of potash stars require different methods. It is a singular circumstance that, though chlorate of potash stars are much fiercer than nitrate of potash stars, yet the latter light without any trouble, while the former, if made in the same way, would be almost sure to miss.

To Make Cut Stars.

Wet the composition with thin starch, or dextrine solution, or gum water, sufficiently to bind; press it into a flat mass, on a slate, or Dutch tile, with a knife, or small trowel, till about $\frac{3}{8}$ of an inch thick. Indent the surface with the edge of the knife, in parallel lines, about $\frac{3}{8}$ of an inch apart, and cross these with equidistant indentations, at right angles. Set the mass by, to dry gradually. When nearly dry, break it up into little $\frac{3}{8}$ cubes, and lay them out, to dry thoroughly. The broken edges will be rough, and will catch easily.

Dry Pill-box Stars.

Take a sheet of note paper, and cut it into four equal parts; each part will be about $4\frac{1}{2}$ inches by $3\frac{1}{2}$. Paste and roll them on a $3\frac{1}{2}/8$ brass tube, so as to have the cases $4\frac{1}{2}$ inches long. To make these into pill-boxes, perfectly true, like those used by the druggist, they must be cut in the lathe. For this purpose, turn a cylindrical piece of wood, fig. 27, which is to fit easily into the case, except at a, where it is to be turned sloping a little larger, so that when the case is slipped over it, it will bite at the part a; otherwise, on putting the chisel to it, it would slip round, without getting cut. Mark the case, with a black-lead pencil, at every half-inch; suspend it in the lathe; and cut it at the marks. Next procure a $\frac{7}{16}$ inch punch, with which to cut out the bottoms. These are to be made of card, or bristol-board. Lay the card on a piece of sheet-lead, or the grain end of a piece of beech, and give it a smart blow or two with a hammer; keep on punching till the punch contains a dozen or more discs, then push them out. When a number are ready, press them into the pill-boxes, with a rammer that fits loosely.

The composition is to be put into these dry, and driven in with a solid rammer, and the little mallet, before described.

This was the original way of making them, but is perfectly unnecessary. Roll the tubes as directed, of two thicknesses of paper, with a little bit to lap over. Cut them across, with one sharp clip, with a strong pair of scissors. This will slightly flatten them; but they may readily be restored to the cylindrical form, by slipping them on a piece of wood, and rounding them to shape with the fingers. No bottoms need be provided, no punch used.

To Fill the Boxes with Dry Colour.

Rest the box on a flat surface, put in some composition, and drive it in with a brass or boxwood drift and the little mallet, till half full, as in fig. 28. Then fill up the box with more colour, set a little bit of match upright in the side, and mallet it in, till the box is almost full. The drift for this must be cut flat on one side, to allow for the match. On the top put a very little dry meal powder, or shell-fuse, and press it in with the finger. Cut a piece of double-crown, about an inch broad, and long enough to go rather more than once round the pill-box: paste it all over; lay the pill-box on it, and roll it up; tuck in one end, to make a bottom, and press the other end round the match, and on the meal powder, or shell-fuse, till it assumes the form of fig. 29.

Another Way.

Set the pill-box on a flat surface, put in a very little meal powder or shell-fuse, then some composition, and mallet it in till full. Roll this up in a piece of double-crown as before; tuck in the bottom, and set by to dry; when dry, put into the other end—the end containing the film of meal or shell-fuse—a piece or two of thin match, and tie it in, as in fig. 30. Dry pill-boxes are best for making chameleon stars; these are half one colour and half another. Make a dozen stars, half yellow and half green, and a dozen more half blue and half crimson; put these into a rocket; they will burst green and crimson, and change to blue and yellow.

Bottomless Pill-boxes.

The cases are the same as before. To fill them, damp the composition as for Roman candle stars; put a bit of quickmatch into the case, as at fig. 31, and press in the composition. This is usually done with the fingers, but is not very pleasant work, especially with lac solution. A cleaner way, though more tedious, is to fit the case on to a little foot, with a side notch in it, fig. 60; then slightly mallet in the composition. Or a notch may be cut in the side of the box, fig. 61, and the match put in, as fig. 62, and slightly malleted.

Instead of making the cases entirely of writing paper, they may be made half of writing paper and half of coloured double-crown, to indicate the colour of the star.

Another way is to sift a thin layer of French chalk over a sheet of paper, and to roll the stars in it, one by one, as they are punched. When dry, brush off the superfluous chalk, and prime with a bit of match, tied across the mouth.

Enveloped Stars.

Pump and drive the stars exactly as for roman candle stars, They may be $3\frac{1}{2}/8$ diameter, and $\frac{5}{8}$ long; or they may be formed with figs. 1, 2, and 3, and the side pin of fig. 2 may be removed to c, in which case the stars will be $\frac{5}{8}$ diameter, and $3\frac{1}{2}/8$ deep; the former will be an oblong cylinder, the latter an oblate. Cut a strip of red, blue, green, or yellow double-crown, of a suitable breadth, and long enough to go twice round the star. Paste the strip all over, or gum it at the edge only, and lay the star upon it, as at a, fig. 41, with a bit of match behind it; then roll it up and put by to dry. When pasted, it shrinks and holds the match tight; when gummed at the edge this is not always the case; it may then have a piece of thin binding-wire twisted round it. It will have the appearance of fig. 32. In fig. 31, the match is embedded, and dries in the damped composition; in the enveloped stars the stars are dried first, and the match lies outside and blows through. Enveloped stars show well in the air.

The word envelope, as used in these pages, must not be confounded with the same word as applied to the coverings for letters. It is rather synonymous with the term wrapper; you lay an ounce of tobacco on a piece of paper, roll it up and tuck in the ends. So with cases: you roll them up in a piece of paper, and leave an inch vacant at each end to receive quickmatch, &c. This is termed the envelope, that is, the wrapper.

Rocket Heads.

Heads for small rockets may be made of two or three rounds of paper rolled dry, and secured on the inner and outer edges with paste. After sticking it on the case, which it may be made to fit, as in fig. 36, pinch the top in like a choke, only tight, and tie it round with twine or flax. For coloured rocket heads, $6/8$ and upwards, the head may be enlarged by fixing a collar round the top of the case. To make the collar for a $6/8$, roll a case on a $9/8$ former, and when dry cut it into short lengths in the lathe, as recommended for pill-boxes; if unprovided with a lathe, saw it with a fine-toothed saw. The advantage of the enlarged head is, that it brings the stars nearer to the rocket, and prevents it from being top-heavy. Another way is to make the heads tapering, and the tops conical, as in fig. 39. One part is rolled on fig. 34, the other on fig. 35; or, a cone may be made of a circular piece of paper, without a former. Cut the circle, along the radius, to the centre; bend it into a cone; secure the edge with sealing-wax, and paste paper over it to overlap the rim: snip the edge with the scissors; paste it inside, and secure it to the other part. But the quickest way of all is to make a long paper bag, which may be made to fit to the greatest nicety. For this purpose take a tape-measure; or lay down, on the edge of a strip of writing paper, 7 or 8 inches long, and 1 inch broad, 6 inches, divided into eighths, transferred from a foot-rule. Suppose the $6/8$ rocket has a collar, which, on being measured by the paper just alluded to, is found to be $4^{2}/_{8}$ inches round; add to this $3/8$ for lapping over, making $4^{5}/_{8}$ inches. Cut a piece of imperial brown $4^{5}/_{8}$ broad, and as long as the case, 6 inches. Make this into a paper bag, $2^{1}/_{8}$ inches broad. Be careful that the corners are perfect; a strip of double-crown may be pasted over them. When dry, pinch the mouth open till cylindrical, and merely allow the stars to drop in without forcing them; this will keep the top edge of a wedge shape, fig. 37, and answer the purpose of a cone. If the paper is thin, the bag must be made of two thicknesses; the paper will, then, require to be 9 inches by 6; $4\frac{3}{8}$ of this will have to be kept dry, and the other $4\frac{5}{8}$ pasted.

A head, made in this way, if required to hold gold rains, or serpents, can instantly be reduced to a cylindrical shape, by pushing the rocket-case right up it, to the top; this will cause the corners to stick out, like two horns; press them down, and secure them with sealing-wax. Attach a cone, if desired.

Heads made like figs. 38 and 39 may be of three thicknesses of paper, pasted all over.

Fig. 39 shows the manner of tying on the sticks. No variation must be made in their lengths, and it is not advisable to alter their size. To adapt it to the wood, however, a slight alteration might be permitted. For instance, instead of $\frac{3}{8}$ square, it might be $2\frac{1}{2}/8$ by $3\frac{1}{2}/8$, a slight increase one way, compensated by a slight diminution the other. It must, however, on no account, be so increased and diminished, as to approach the shape of a lath, as such stick would vibrate, and cause the rocket to quiver. When the heads are a paper bag of the shape of fig. 37, the stick must be tied on, as indicated by the dotted lines.

If the stick is suited to the rocket, it will, when suspended on the finger almost against the mouth, as at f, fig. 40, lie, not quite horizontal, but slightly sloping downward. The wood should be dry pine, free from knots. The sticks are generally cut with a carpenter's cutting-gauge. If the learner has a lathe he will find a 6-inch circular saw convenient for cutting them.

For amateurs, a $\frac{6}{8}$ rocket is a good size; large enough, and small enough. If $\frac{5}{8}$ and $\frac{4}{8}$ are made, two hollow drifts will be sufficient; for $\frac{3}{8}$, one hollow drift. Very small rockets, $\frac{2}{8}$, are made for children; they are rammed solid, and a hole is pushed up them with a bradawl.

Rockets in former times, before the present days of competition, were charged in moulds. These were of gun-metal, bored truly cylindrical, furnished with hinges, to open and admit the case; they were then screwed up, and might be charged as hard as possible. Names were given them according to the bore of the mould, that is, the external diameter of the cases; a ⅝ was termed an ounce rocket; a ⅝, a two ounce; a ⁶⁄₈ a quarter pound; a 7½/8 a half pound; a ⁹⁄₈, a pound; a ¹²⁄₈, a two pound.

These names were determined by the weight of a leaden ball of the same diameter as the bore. Now a sphere of lead, 7½ inches diameter = 90 lbs., or 1440 ounces; consequently one of 15 inches = 720 lbs., similar solids being to each other as the cubes of their like dimensions: the latter sphere being twice the diameter from top to bottom; twice the diameter from left to right; and twice the diameter from front to back; $2 \times 2 \times 2 = 8$.

The external diameter of the rocket being 7½ inches, the internal would be 5 inches, and this would be a 90-pounder: hence, for an inch rocket, we have the proportion,

$5^3 : 1440 \text{oz.} :: 1^3 : 11 \cdot 52 \text{oz.}$

so that a 12-ounce, or 3-quarter-pound rocket, ought to be a trifle above an inch.

Required the weight of a ⁶⁄₈ rocket. 5 inches = ⁴⁰⁄₈.

$40^3 : 1440 \text{oz.} :: 6^3 : 4 \cdot 86 \text{oz.}$

So that a ⁶⁄₈ is a trifle too large for a quarter-pound.

If, conversely, we require to know the size of a half-pound, or 8-ounce rocket,

$1440 \text{oz.} : 40^3 :: 8 \text{oz.} : 3200/9$

and $\sqrt[3]{(3200/9)} = \sqrt[3]{(9600/27)} = (\sqrt[3]{9600})/3 = 7 \cdot 08$

So, properly, a half-pounder is a trifle over ⅞. The names in use enable the makers to understand each other, but they are not mathematically correct, and are of no utility to an amateur.

The following table shows the true weight of leaden spheres, the dimensions being taken in inches.

Weight.	Diameter.	Weight.	Diameter.	Weight.	Diameter.
1 dram	·264	1 oz.	·664	1 lb.	1·672
2 drams	·332	2 "	·836	1¼ "	1·8
4 "	·418	4 "	1·056	1½ "	1·91
8 "	·528	8 "	1·328	1¾ "	2·02
12 "	·604	12 "	1·52	2 "	2·112

And ⅔ of the above numbers multiplied by 8, give the correct names for rockets in eighths of an inch.

Required the true size of a half-pounder.

(1·328 × 2 × 8)/3 = 7·08.

A trifle above ⅞, as before stated.

A cast-iron ball 6 inches in diameter weighs 30lb.

Cast iron is about ⁴⁰⁄₆₃ the weight of lead.

5280 feet = 1 mile; 3280 feet = 1 kilometre.

WHEEL AND FIXED CASES.

Wheel cases may be $5\frac{5}{8}$ inches long. Cut the paper, without waste, into 4 strips, each 29 inches long. Fixed cases may be $7\frac{1}{4}$ inches long: cut the paper the other way of the sheet, into 4 strips, each $22\frac{1}{2}$ inches long. The extra length, for fixed cases, is to allow of their being reported; that is, filled at the end, with an inch, or more, of grain powder, to make a bang like a squib. The cases are to be gauged to the thickness of roman candle cases, but choked like rockets; $\frac{5}{8}$ internal, and $\frac{7}{8}$ external, are a good size. Two solid drifts will be required for wheel cases; one $5\frac{1}{2}$ inches long, besides the head; the other 3 inches: for the fixed cases it will be necessary to have one of $7\frac{1}{4}$ inches. They may be turned of box, beech, or ash; but gun-metal drifts are best, though by no means indispensable. If they are cast in gun-metal, they will simply require filing in the rough places. The cases are to be rammed solid throughout, on a foot, fig. 42, turned in one piece, with a nipple, and a pin $\frac{1}{3}$ of a diameter thick, and just high enough to keep the choke clear.

In all wheel and fixed cases, whatever the remainder of the fuse may be, begin by putting in one scoop of starting fire; this, when malleted firmly in, should fill about $\frac{3}{8}$ of an inch in the case. The subsequent fuse can be selected from the Tables. The choke may be protected with a little clay, before the starting fire, like rockets, if thought desirable: no clay is to be used anywhere else. Wheel cases containing steel-filings are termed brilliant.

To prime the cases, very slightly damp some meal powder, by sprinkling it with a few drops of water; mix and chop it up, or mince it, as it were, with a knife: put a little into the mouth of the case, press the nipple, fig. 42, into it, and work it round; this will prime the choke and the mouth at once, still leaving the choke clear; or, paint the choke and mouth with the sash-tool, with meal paste, just sufficient to wet them, then plunge them into dry meal, give them a rap to shake off the superfluous dust, and lay them by to dry.

Wheel and fixed cases need not be choked at all, but plugged $\frac{1}{4}$ of an inch, with plaster of paris, as directed for roman candles. When dry, charge them by setting them flat on the block, without using the foot. Mallet in a scoop of starting fire, then the other. Afterwards, when a number are charged, bore through the centre of the plaster, with a shell-bit, ⅓ the diameter of the case; insert a piece of match in the hole; wash the face of the plaster over with meal paste, and plunge into dry meal. A sheet of double-crown makes 8 wheel case envelopes (2.2.2) 10 by 7½, so that the void at each end, to receive the match, is nearly 1 inch.

It is not possible to devise a formula that will indicate the exact quantity required for wheel cases, as the fuses vary; but, representing the length, in inches, by I; and the diameter in eighths by e, the following will help to serve as a guide.

$(I\ e^2)/7$ = drams.

Suppose a wheel case 5⅝ inches long, and 4 eighths diameter,

$(5⅝ \times 4 \times 4)/7 = {}^{90}/_7 = 13$ drams.

Suppose a fixed case 7¼ inches long, and 5 eighths diameter,

$(7¼ \times 5 \times 5)/7 = {}^{181}/_7 = 26$ drms. = 1 oz. 10 drms.

GERBES.

Gerbes, so called from the French word for wheat-sheaf, which they resemble, are fixed choked cases: they do not show well on wheels. As they contain grains of iron, they must not be under $6/8$; for private exhibitions a good size is $9/8$, $7/8$ internal diameter; $12/8$ external; $3/8$ the diameter of the choke; $11\frac{1}{4}$ inches, length of case. Charge the case on a nipple, exactly like a wheel case. It is advisable to put in, first, a little clay, to protect the choke, as the fire, being fierce, would, otherwise, enlarge it, and diminish the ascent of the sparks. Upon the clay drive in a scoop of starting fire, and fill up with gerbe composition. This, when containing iron borings, is termed Chinese fire; the pieces, Chinese trees.

The most magnificent of all, however, is the coloured gerbe. For this, some green, blue, and crimson grains, or small stars must be prepared. They may be quarter-inch cubes, cut as directed for the chopped nitre stars. A far better way, however, of preparing them, is to remove the pin a, of fig. 2, up to c, so that the stars, driven in the tube, will be $5/8$ inch diameter, and $3/8$ thick. When these are dry, chop them into 4 pieces, by holding a knife, or chisel, across them, and giving it a smart blow with the mallet. They are harder made this way.

To charge the cases: having driven in the clay, and the starting fire, put in 7 or 8 stars, then a scoopful of fuse No. 1 or 2, then 7 or 8 more stars, and another scoop of fuse; mallet the whole 4 layers down firm, with blows not too heavy at a time, but many times repeated. Then put in 7 or 8 more stars, another scoopful, 7 or 8 more stars, and another scoop, and mallet the 4 layers as before;

and so repeat. The two layers and the two scoopfuls may fill up, when malleted in, $\frac{3}{4}$ of an inch, in a $\frac{9}{8}$ case.

Instead of choking the cases, plaster of paris is far preferable. Let it be $\frac{3}{8}$ of an inch thick, and well dried, before charging. After the cases are charged, bore a hole through the plaster, $\frac{1}{3}$ of a diameter, that is, with a $\frac{9}{8}$ case, $\frac{3}{8}$ of an inch diameter. For this purpose it is not necessary to have another shell-bit; bore it with the $\frac{3}{16}$, and enlarge it with a penknife. It is better arched under, till conical, as shown in fig. 48. Prime with 4 or 5 pieces of match, and wash with the sash-tool.

The gerbe being finished, make a cylindrical box, or paper bag, of 2 or 3 thicknesses of paper: fill it with a number of crackers, and a scoopful of meal powder, and fasten it to the gerbe.

No single piece is more effective than a coloured gerbe; the stars will be projected 30 feet, or more: they may be put in, mixed; or, one layer may be blue, another green, another crimson. Twist a piece of wire, deprived of its elasticity, round the neck, and another piece round the bottom, and leave long ends; it can then be fastened by them to the top of a post. To remove the elasticity from iron wire, lay it in the fire till red hot; withdraw it with the tongs, and put it aside, to cool slowly. If copper wire is used, it will bend without preparation. String must not be employed, as it might burn, and let the case fall. Common pins, patent short whites, deprived of their elasticity, are useful for connecting the parts of lustres together.

FLOWER POTS.

These are choked cases, charged with spur fire: the fire is somewhat slow, so the cases must be short: 4 inches long, and $\frac{5}{8}$ diameter is a good size. Rub the composition thoroughly up in the mortar; the vegetable black produces beautiful star-like sparks, totally dissimilar to any other. Put a little composition, at a time, into the case, and jolt it with the roman candle rammer.

Vegetable black, introduced into a star, causes it to tail, like linseed oil. Light such star on the hob; it will burn, and leave a residue, unaltered in shape; blow upon this continuously with the mouth, or, better still, with a pair of bellows: the supply of oxygen will cause it to boil up, in a state of fusion, when it will begin to throw out clusters of the peculiar starlike sparks, before mentioned, bright and yellow as new sovereigns.

Vegetable black is a pure lamp black; some samples of lamp black make equally good stars, but others are worthless. Greater reliance can be placed upon vegetable black. Vegetable black and lamp black must not be mixed with linseed oil, as such mixture is liable to spontaneous combustion.

Roll up a tube for pill boxes, of two thicknesses of brown paper. When dry, cut it into pieces about $1\frac{1}{8}$ inch long: choke one end, like a wheel case; set it on a nipple, and charge it with spur fire, till full within $\frac{1}{16}$ of an inch: fill up flush with a little plaster of paris, pressed in flat with a knife: prime the choked end, and put a number of such cases into a rocket head, or shell.

PORT FIRES AND SHELL FUSES.

These are unchoked cases, like roman candles; 6 inches long, $3/8$ internal, $4\frac{1}{2}/8$ external, is a good size. They should be rammed, as hard as possible; and, for this purpose, it is best to have a mould. Now, if a case is rolled of such a size that it will exactly fit into a brass tube, and is charged, in it, very hard, it will swell, and it will be almost impossible to get it out again; but if it be made a trifle smaller, so as to just slip through the tube; then, if a piece of writing paper be rolled, dry, round it, once or twice, so as to make it a tight fit, and the case is charged, it can be pushed out, like a pellet from a popgun, leaving the writing paper, generally, in the tube, or mould, and the case will come out without a wrinkle.

Let the composition be put in, very little at a time, and well driven with a solid rammer and mallet. Fig. 59 represents the mould; the foot b fits the tube a; the tenon c fits the case; a wire, d, goes through 2 holes in the brass tube, and a hole through the foot; a nut, e, to keep the wire from jarring out, is made of a piece of indiarubber: make a hole through it, with a bradawl, and slip it on the wire; or, a screw-eye may be passed through and held with a leaden, or wooden nut.

TOURBILLIONS.[A]

A tourbillion, so called from the French word for whirlwind, is a case made to rotate and ascend at the same time, forming a spiral of fire, and ending in the shape of an umbrella.

To Make a Tourbillion.

Roll the case like a roman candle case, but gauge it to the thickness of a rocket case. Let the inner diameter be $6/8$; the outer $9/8$; the length of the case $7¼$ inches, fig. 43. To charge the case, have a mould, as directed for port-fires; and let the tenon rise exactly $3/8$ of an inch up the case. Put in a little composition at a time, and mallet it as firmly as possible, till within exactly $3/8$ of an inch of the top of the case; so that there will be a vacancy of $3/8$ of an inch, at each end. Fill each of these ends flush with plaster of paris. It is, better, too, if you can manage to fill the middle half-inch of the case with plaster of paris. It can be effected with care, and will hold the screw, hereafter to be described, more firmly.

Construct a wooden box, fig. 44, consisting of a bottom and two sides only, firmly screwed together. Each of the pieces of wood is to be $7¼$ inches long, and $½$ an inch thick. The internal breadth of the box is to be exactly $9/8$ of an inch; and its internal depth exactly $4½/8$ or $9/16$, so that when the tourbillion is laid evenly in it, and pressed down to the bottom, half of the case will be in it, and half out of it. At a point b, fig. 44, on the top of the side, half an inch from a, make an ink mark: and, at a point d, half an inch from c, make another ink mark. Fig. 45 is the bottom of the box. At a point w, $5/8$ of an inch from the end; and, at a point z, $5/8$ of an inch from the other end, make holes with a fine bradawl, truly, in a line down the middle of the wood, as between side and side. The distance w z is 6 inches; divide it into 3 equal parts, in the points x and y, two inches asunder. Bisect x y in the point s. Procure 5 carpet pins, fig. 47: they will, probably, be $¾$ of an inch long. Drive them through the holes w, x, s, y, z, inverting the box for the purpose, so that they shall stand bolt upright in the box. Now screw, or nail a piece of wood over the bottom of the box, entirely to cover it, to prevent the carpet pins from getting displaced. It will be seen, that, if the tourbillion be now laid evenly in the box, and pressed down till it rests on the bottom, the projecting pins will make 5 holes, in the under part of the case. While it is thus lying, with a stiletto, such as used by sempstresses, for making eyelet holes, prick the side of the case over the line b, of fig. 44; and, also, over the point d. There will now be 7 holes; 5 underneath, 1 to the right, and 1 to the left: the latter are the places for the whirlers, or holes of rotation: 4 underneath, for the lifters, or holes of ascension; the centre one, s, receives a nail or screw. Take a bradawl, fig. 49, $\frac{3}{16}$ of an inch diameter; and slip over it a shield, consisting of a piece of wood with a central hole up it, like a pop-gun, of such a length, that, when it is slipped on, only $\frac{1}{4}$ of an inch of the bradawl protrudes; or, instead of a bradawl, fix in a handle, a wire of equal length, namely $\frac{1}{4}$ of an inch, and file it to a point. Push this into all the holes, except s, making 4 holes underneath, and 2 horizontal holes, one left, one right: all these holes will be exactly of the same depth, on account of the shield: see that they are bored perfectly true, the horizontals exactly 90 degrees above the others, or $\frac{1}{4}$ of the circumference.

The next thing required, is a piece of hooping, or curved stick, about $\frac{6}{8}$ of an inch broad, and as long as the case, $7\frac{1}{4}$ inches. In the centre of this, bore a hole, and countersink it; then, with a screw, an inch and a quarter long, screw the hooping, at right angles, on the bottom of the case, through the point s of fig. 45, which must be enlarged to receive the screw. A touch of glue may still farther hold the wood in position. It will now assume the shape of a cross, like figs. 50 and 51. Fig. 50 shows the under side of the case; fig. 51 the upper. From w to x lead a bit of naked match; push the ends into both holes, and secure with a little wetted meal, pressed in with a knife. Do the same with y and z. Paste a piece of double-crown, $\frac{3}{4}$ of an inch broad, and of sufficient length, and cover each of the two pieces of match, with two layers of the paper. Turn it over, like to fig. 51; connect the holes a and b with a bit of naked match; and, under the centre of it, slip another piece of naked match, having a piece of touch-paper round the protruding end: cover the match with two thicknesses of pasted paper, in the same manner as the under holes. The tourbillion is now complete. See that it will balance, and swing round easily, when laid upon a level surface. The proper way to fire it, is from a flat sheet of iron, or a flagstone. Light the touch paper; the fire will communicate to the side holes, and set it in rotation. As soon as $\frac{1}{8}$ of an inch of fuse has burnt from each end, and the piece has got well into action, the 4 under holes will catch, and cause it to ascend.

Instead of this mode of making a tourbillion, some charge it with an inch of solidly rammed clay, in the middle; fasten the stick, by crossing it with binding-wire; bore a hole through the middle of it, and of the clay, and slip it over a tapering-wire, standing upright in a block, like the spindle of a rocket. Four holes only are then used; two of rotation, and two of ascension; and the whole are fired at once, the match starting from one of the under holes, going to the side hole; over, across, to the other side hole, and on to the other under hole.

Instead of making them with clay in the middle, there is yet a better method of having two cases, each about 4 inches long; and gluing, or fastening them with tin-tacks on a centre-piece, turned with a tenon at each end, fig. 52; two balancing arms, one on each side, must then be fixed to the centre-piece.

In a windmill, as is well known, the vanes, or sails, are set at an angle. There is a toy, made of two slips of tin, forming a cross, and set at an angle, sloping upwards, called the flying dutchman; this, when spun with a string, from a handle like a humming-top, flies up into the air, on escaping from the string. Steel Fliers, with two vanes, are used by sportsmen to practise shooting flying. Small balloons, some years ago, in a room in the Polytechnic Institution, free from a current of air, were guided or driven by a similar contrivance, moved by clockwork. The screw-propeller of a ship acts on the same principle. I think it possible that, if two vanes were fixed in the central piece of wood, set at an upward angle of 10, 15, or 20 degrees from the horizontal, they might assist the ascension, and so cause the tourbillion to reach a greater height; or, the vanes alone might cause it to rise, upon 4 side holes, two to the left, and two to the right, causing it to rotate. The design is shown at fig. 54, the shape of the vane at fig. 53. I have not yet tried it, so offer it only as a suggestion.

FOOTNOTES:

[A] Tourbillion, from *tourbillon*, like postillion, from *postillon*, the i being inserted to approximate the pronunciation of the French. In pavilion, from *pavillon*, and vermilion, from *vermillon*, one l is dropped; so in battalion, from *bataillon*; while medallion, from *medaillon*, retains the ll.

SAXONS.

These are unchoked cases, charged like a tourbillion, but pierced only with holes of rotation, for the purpose of turning a coloured fire. Drive them in a mould, as directed before. A good size is, $5/8$ internal, $7/8$ external, 6 or 7 inches long. Let the tenon enter the case $3/8$ of an inch: charge the composition firm till within $1/2$ an inch of the top, which leave vacant. Remove it, and fill the $3/8$ occupied by the tenon, with plaster of paris. Have a centre-piece, turned like fig. 55, with a tenon, $5/8$ diameter at each end, $1/2$ an inch long. Glue a case on each tenon. Let the centre-piece be 6 inches long, exclusive of the tenons; so, if the saxon cases are 7 inches long, each, the entire length, as fig. 56, will be 20 inches. Make a hole at a, and another at c, with a shielded bradawl, $3/16$ of an inch diameter. Put a bit of naked match in the hole a, carry it round x and y, along to c and on to z. It must be pushed into c with a blunt wire. Cover it with two thicknesses of pasted paper, like the tourbillion. Leave the match exposed at x, y, and at z, and brush it over with meal paste. If the central piece of wood were now put on a horizontal spindle, and fire communicated to the match z, the holes a and c would cause it to rotate, and produce a white circle of fire. This, however, would be hardly worth making; but, by fixing at b, a little case of coloured fire, a splendid effect is produced. This case of colour is usually tied to a nail, driven in at b; but a little tenon of wood may be glued there instead, and the case of colour must be then charged, with a vacancy at the bottom, to fit on the tenon. The case of colour must be timed to burn as long as the saxon; rather more than an inch will be sufficient.

Saxons are sometimes made by charging a roman candle case with an inch of clay in the middle, and boring a hole through the clay, to receive a spindle. Only one half of the case burns at a time; a leader, placed at the bottom, near the central clay, conveys the fire to the other end; and continues the rotation. For distinction, they are called Chinese fliers.

FIVE-POINTED STARS.

These are cases about 2½ inches long, and 1 inch diameter. Make a bottom to the case with ¼ inch thickness of plaster of paris, so that it looks like a large pill box. Charge it solid, and at ⅜ of an inch from the extremity, that is, ⅛ of an inch beyond the plaster bottom, round the circumference make five holes, as for saxons; run a bit of match round, connecting the holes. These, when fired, stand out at right angles, the plaster towards the spectator, so that the fire resembles a gas star, with 5 points, as in fig. 130.

SQUIBS.

Take a sheet of 60-lb. imperial brown, and a sheet of 12-lb. white demy. The imperial, as said before, is 29 by $22\frac{1}{2}$. Cut it into 24 equal parts (2, 4, 3), that is, first into two equal parts, down the natural fold of the paper; then each into four equal parts, at right angles to the first fold; and each of these into three equal parts, at right angles to the second fold. Each piece will then be $5\frac{5}{8}$ by $4\frac{5}{8}$. The demy is $22\frac{1}{2}$ by $17\frac{1}{2}$. Cut the sheet into sixteen equal parts (2, 4, 2), each piece will be $5\frac{5}{8}$ by $4\frac{3}{8}$. A piece of brown and a piece of white will make a case $5\frac{5}{8}$ inches long. For a former, procure a piece of brass wire, or stair-rod, about a foot long, and $\frac{1}{4}$ inch diameter. Lay eight pieces of the demy evenly on each other; draw the thumbnail of the right hand from the farther edge of the paper straight over the middle towards you, a few times. If properly performed, this will draw piece behind piece; proceed till about $\frac{1}{8}$ of an inch of each is left exposed, in the same manner as a pack of cards would arrange themselves, if set upright, and allowed to fall forwards; something after the manner of the laths of a venetian blind, or slates upon a roof, imbricated. If you cannot acquire the knack of doing this, you must so lay them, one by one. Paste the edges of all the eight pieces thus lying. Place one of the pieces of brown paper before you; lay the former, or stair-rod, across it, nearly in the middle; bend the paper over it, and press it in with the fingers of both hands; roll for an inch, or so; lay it on the middle of a white piece; bend the white over; infold the brown in it, and roll forward, till it catches the pasted part, and sticks. After a number have been prepared, and are dry, choke them, as directed for rockets. The wire of the choker, fig. 24, should be about $\frac{1}{16}$ of an inch thick. Now take a piece of square steel, or iron wire, 12 inches long, the thickness of the wire being about $\frac{3}{5}$ the diameter of the stair-rod, that is, $\frac{1}{4} \times \frac{3}{5} = \frac{3}{20}$ of an inch; if it be found difficult to procure square wire, file a round piece. Bore a hole down a bradawl or chisel handle, 2 inches in depth, and $\frac{1}{4}$ diameter; fix the square wire in it, with melted lead, like fig. 4; the lead is for the purpose of giving weight to the blows in charging. Black the wire all over with ink, and allow it to dry. The next requisite is a tin funnel, without a neck, of the size and shape of fig. 65. Any tinman will readily make one to order; but if the learner procure a soldering-tool, he can construct one himself. For this purpose, dissolve a piece of zinc in a little hydrochloric or muriatic acid, till the acid is saturated: heat the tool, and dip the tip end, momently, into it; the acid combines with the oxide of copper formed by heating, and the zinc adheres to the clear surface of copper produced; it will now easily take up the solder: the joint to be soldered must be clean, and also touched with a feather dipped into the acid. It is best to make a funnel with a piece of writing-paper, first, for a pattern; when this is got correct, the tin can be cut according to it. By describing a circle with a 3-inch radius, and cutting out a sector of 100°, the correct size is obtained at once; $\frac{1}{8}$ of an inch breadth, outside the radius, is to be allowed to lap over. The bore of the squib is $\frac{5}{20}$ of an inch; the size of the ramming wire $\frac{3}{20}$; the hole at the bottom of the funnel may be half way between, $\frac{4}{20}$ or $\frac{1}{5}$ of an inch diameter. Push the wire rammer through the funnel, till it protrudes 2 inches beyond the bottom; observe the part of the wire which is now level with the top of the funnel; withdraw it, and file a bright mark round the part: it will be about $4\frac{1}{2}$ inches from the end. The wire, having been inked, shows the bright mark more plainly.

Make a deal box, 3 inches square, and 4 inches deep, and nail a bottom to it. Also have a little nipple, fig. 73, with a wire, as drawn; it can be secured to a flat board.

To charge the cases. Set the choked end of one on the nipple, insert the funnel in the other end, put the wire rammer down through the funnel, and let it fall to the bottom of the case: put in some fuse, and jolt the rammer up and down, till the case gets so far filled, that the bright, filed notch, before described, is on a level with the top of the funnel. It is necessary to sit at a table to charge the cases; the eye is, then, on a proper level to see the mark. As the cases are thus charged, set them in the square box, choked end downwards, till a number are filled. Now take out a handful, invert them over a sheet of paper, and give them a few taps with the rammer; this will shake out a great deal of loose composition, that has got puffed up, in the case. Return them to the deal box.

The next operation is to bounce, or bang them. For this purpose take a long slip of paper, 7 inches broad. Stretch it lengthwise before you. Lay a number of the cases upon it, so that all the choke ends lie flush with the left edge of the paper; then roll them up in the paper; turn and set them upright on the choke ends; the paper will now stand up more than an inch above the cases, as a tumbler, three parts full of water, stands up with an empty space above the water. Rest them on a sheet of paper, and pour in a quantity of F grain powder; this will fill every case; loosen the paper wrapper, and allow the superfluous grain powder to fall on the paper. Set the cases, choke downwards, in the square box. As they are all full, it is necessary to get a little out of each. For this purpose, lay the square box, containing them, horizontally on a sheet of paper, on one of its sides; turn it gently over, and lay it on the next side; a little of the gunpowder will spill out; turn it gently over again, and lay it on the next side; and so proceed, till every case has about $\frac{3}{8}$ of an inch empty. The ends of the cases are now to be closed. Take a case, in the left hand, as in fig. 63, and wind a piece of string, or whipcord, three or four times round it, holding the end of the string, a, firmly with the thumb; then, with the right hand, bring the end, b, back over all the folds, as in fig. 64; pull the end, B, tight, and the case will be closed.

Melt some common bottle sealing-wax in a pipkin; carry it to a distance from the fire, and dip the ends of the squibs into it. Next, prime them, by pressing the choked ends into very slightly damped meal powder, as directed for wheel cases. Take a piece of touch-paper, 15 inches by 10. Divide it into 48 pieces (2, 2, 3, 4); each piece will be $2\frac{1}{2}$ by $1\frac{1}{4}$. Roll a piece round the primed end, twist to a point, and fasten it with a bit of carpet thread, hitched three times round it.

SERPENTS.

These are simply squibs, made short, in order to burn out quickly, as they are intended to be thrown from mines, and would otherwise lie too long on the ground. They are best made entirely of brown paper, pasted all over, exactly like wheel cases, and choked in the same manner. Cut the paper 6 inches by $2\frac{1}{2}$, and roll them on the squib former, so that they shall be $2\frac{1}{2}$ inches long. Ram them with the funnel and wire, but continue the ramming till the guide-mark stands a little higher than the level of the funnel; until, in fact, the cases get half full. Bounce, close, and dip them in melted wax, like squibs. The best way to prime them is to paint them with the sash-tool, slightly, then plunge them into dry meal, so that they may be, as nearly as possible, like quickmatch. If they were pressed into wet meal powder, of the consistency of mortar, and then rubbed smooth on a Dutch tile, or slate, or plate, they would dry hard and glazed. In this case, if they were fired from a mine, in all probability three-fourths of them would miss. In brushing them with the sash-tool, with meal paste, turn the brush well round, to leave little in; so that, after being pressed into the dry meal, they may not be choked up, but the cup shape still be left. They are not to be touch-papered, being intended only for mines, rockets, or shells.

PINWHEELS.

Procure some 16-lb. double-crown white paper.

A sheet is	30 × 20 inches.
Cut off a strip	30 × 3
Leaves a piece	30 × 17

The small strip may be used for odd purposes. Divide the 30 by 17 into 10 strips, each 3 by 17, for penny pinwheels. If the paper were cut the other way of the sheet, it would not wind smoothly, but crumple up. The fibres, from some cause or other, appear to arrange themselves in one direction, like the grain in wood. For halfpenny wheels, divide the sheet into two pieces, each 10 by 30; then cut each of these into 12, each 10 by 2½. For a former, have a straight piece of iron, or steel, wire, ⅛ of an inch thick, and 24 inches long. One end of this must have a basil, or sloping enlargement upon it, which is thus made. Take a piece of double-crown 4 inches square; lay it straight before you, and cut it diagonally, from the right top corner to the left bottom corner; remove the left half; paste the right; lay the wire upon it, so that the rectangular corner is 5 inches from the right end of the wire; roll it up, and press it smooth; the sloping edge will now form a spiral, commencing at 9 inches from the right end of the wire, and being enlarged at 5 inches from the end. Paste a thin strip of paper over the whole, to protect the spiral edges. It will now be something of the shape of fig. 27, only turned the reverse way. The 5 inches at the end are for a handle. To roll the pinwheel pipes, lay the strips of paper evenly upon each other, and work them back with the thumbnail, as before. Paste the edges; lay half-a-quire of double-crown, or of blotting-paper, flat upon the table, to roll on. Place one of the strips of pasted paper upon it; lay the wire upon it, the basil end being towards the right hand; the wire is to be laid, not quite parallel with the near edge of the paper, but slightly sloping, about an inch and a quarter distant from it on the left, and not quite an inch on the right, or basil end. So much of the basil must lie on the paper as will make the mouth of the pipe large enough for the nozzle of the funnel to enter half-an-inch. The basil, of course, gathers up the paper more quickly than the wire, and brings the edge straight, on the completion of the pipe. Now to roll, bend the paper over the left end of the wire, press it in with the fingers, and begin immediately to move it forward; at the same time draw the fingers of the right hand rapidly along, pressing in the paper, and rolling forward, keeping the right, or basil, end of the wire tight, and slightly pulled towards you: the paper round the left end of the wire will have thus become rolled half round the circumference of the wire, before the basil end has begun. There is very great difficulty, at first, in rolling so long a pipe as 17 inches; it is advisable to begin with the halfpenny pipes, which are much shorter; the same wire-former will serve for both. If you cannot succeed, get a person to help you; sixteen fingers can manage the matter easily; but it is best to master the difficulty yourself. Have a funnel as nearly the size and shape of fig. 66 as possible. It is best with a lid soldered on the top, with a circular hole in the middle, about the size of a shilling, to prevent the composition from getting spilt. A slightly tapering neck is also to be soldered on at the bottom. Take especial care that the mouth of the funnel dips into the neck, and not have the neck sticking up in the funnel. The joint inside must be perfectly smooth. A charging wire will now be required; this should be a square steel wire, as large as the neck of the funnel will admit, to move easily up and down; let it also have a piece of lead at the end, to give it weight. Take a piece of brass tube, about $\frac{3}{8}$ of an inch diameter, and an inch long. Hold the wire in it, with a piece of flannel, or woollen cloth, and pour in melted lead. To charge the cases, first bend a quarter of an inch of the small end of the pipe, over the edge of a knife, into a hook, to prevent the composition running out; insert the nozzle of the funnel in the enlarged end; hold the part of the pipe now round it, with the thumb and forefinger of the left hand, slip the wire, through the funnel, down to the bottom of the pipe, which must rest, for a moment, on the table; pour in the composition, jolt the wire up and down; lift the pipe from the table, keep jolting the wire, at the same time turning it round, and let the pipe swing slightly to and fro, till filled. A guide-mark must be filed round the wire, as for squibs, at such a distance as that, when it is level with the top of the funnel, the bottom may protrude about $\frac{3}{8}$ of an inch. When a number of pipes are charged, close the large, or basilled, end, which is the one to be lit, by means of a piece of whipcord, or thin twine, wound round it, as directed in figs. 63 and 64. Soak a towel in water; wring it as dry as possible; spread it out flat; lay the pipes in a row, side by side, upon it, like rushes in a chair bottom; roll them up in it, and leave them for 10 or 15 minutes. Then wind them upon the usual circular blocks, and fasten the end with sealing-wax. Be careful that the wax is not in a flame at the moment of touching the case; if necessary, blow it out. Accidents will arise, both from the wax and from the candle, if care is not used. Measure, with a tape, round the wheel, now wound; suppose 4 inches; cut a piece of blue double-crown, 4 inches broad, and any length: cut this into strips about $3/16$ of an inch broad, and 4 inches long; paste a slate all over; lay these strips, 7 or 8 of them, side by side, flat upon it; paste their upper surface; lay one across each pinwheel, and bend it over to make the ends meet in the centre on the other side. If you wish the pinwheel to be of double or triple size, after you have charged one pipe, paste the small end of a second pipe, outside, and stick it in the other. Let it dry; then charge the additional pipe. Observe, the pipes must not lie too long in the towel; if the nitre gets dissolved, it soaks into the paper, which becomes, virtually, touchpaper, and one pipe ignites the other. The wire formers must not be allowed to get rusty; therefore, before laying them aside smear them with tallow, or olive oil. The wire is much facilitated, too, in delivering, if, just before using it, it is slightly oiled, and then wiped, apparently, perfectly dry. After the wheels are finished, they should be spread out to dry, as soon as possible. It is not usual to prime them; and, as sold, they are very difficult and troublesome to light; for private use, they may have a short piece of match inserted in the mouth, and a small bit of touch-paper wound round it. French nails are now made, of wire, very thin; those, about an inch long, are much better than pins, for firing the wheel upon, as they have larger heads, and prevent its falling off. Pinwheel blocks may be obtained of Merrick, Shuttle Maker, 155, Bethnal Green Road; the same person sells triangular and vertical wheels; caprice, furilona and pigeon frames; rocket sticks; mine bottoms; saxon centres, &c.; also, plait mills, made to order.

CRACKERS.

A good tough paper for these is 24 mill, 19-lb. double-small hand.

A sheet is　　　30 × 20　inches.
Cut off a strip　30 × 4
Leaves a piece　30 × 16

The strip 30 by 4 may be used for squibs, or any odd purpose. For penny crackers cut the 30 by 16 into 8 strips, each 3¾ by 16. For halfpenny crackers divide the sheet into 2 pieces, each 10 by 30. Cut each of these into 10, each 10 by 3. For a former have a straight piece of iron, or steel, wire, 3/20 of an inch diameter, and 24 inches long. This will not require a basil. Have, also, another wire, of half the thickness, for an opening wire. Lay the strips on blotting-paper, as before; and place the wire parallel with the edge of the strip, as there is now no basil. When a number of pipes are rolled and dry, hold one flat on a table; and, with the handle of a knife, or toothbrush, rub the pipe flat along, all except the first half-inch, held in the left hand, to form a little cup. Now push the opening wire through it, to partially open it. Bend the bottom, over a knife, into a little hook, as before directed, for pinwheels. Take up a quantity; wind a strip of paper round them, as in banging squibs, and pour in a quantity of FFF, or canister powder. Unless the powder is very fine it will not run down. Pipes may be filled with F grain powder, one at a time, by putting a funnel into the cup part, and tapping the pipe with a wire. You can hear the powder run down, and easily judge when the pipe is full. The powder is now to be crushed into meal, by rolling the pipes through a plait mill, fig. 71. This is furnished with a wooden screw, at top; the screw presses upon a cross-bit, which bears upon two movable collars: these rest on the axis of the top cylinder, by which means the pressure can be adjusted as desired. The cylinders of plait mills are of beech, or boxwood; the latter are, of course, the best; but for making great numbers of crackers, it is necessary to have the frames of iron, and the cylinders of steel. If a mill cannot be procured, the pipes may be rolled with a brass tube, or rolling-pin, or passed through an American mangle; but a very good, and far less expensive plan, is to lay them on an anvil, or flat-iron turned upside down, and beat them with a smooth-faced hammer; this is a tedious process, but it makes good crackers.

The cup end, having served its purpose, is to be flattened, and bent into a hook. Now lay the pipes in a damp towel, like the pinwheels. Take two pieces of deal, each about 6 inches long and 1½ inch broad: let the one for the halfpenny crackers be ¾ of an inch thick; the other, for the penny, 1 inch thick. Saw and chisel a piece out of each, about 4 inches long and ½ an inch broad, as drawn, fig. 67, so that it looks like the first and third fingers of the hand held straight up, with the second finger, between them, shut. Next procure 20 pieces of wire, each 2 inches long, and about 1/16 of an inch thick. Support the cracker-frame in a vice: lay the cracker across the opening; place a wire upon it, as at a, fig. 68; bend the pipe over it; lay on it another wire, on the other side, as b; bend back, and lay on it the wire c; and so on, alternately, till the cracker is bent up. Lift it out of the frame, and let the wires drop. The distance between the wires, in the halfpenny crackers, will be ¾ of an inch; in the penny, 1 inch; if made according to the directions given. The outside of the frame should be very slightly tapered, or it will be difficult to lift the cracker up out of it. The penny size will probably take 14 wires; the halfpenny 10 wires. Cut the end flush, to expose the crushed powder; wind round it a piece of touch-paper, about 1¾ inch long, 1¼ broad; it need not be pasted; fold up the cracker; pass a piece of flax or thread twice round it; twist the thread in and out, backwards and forwards, among the folds; and the cracker is complete. If the paper does not appear sufficiently thick to make a good report, cut the strips broader, and fewer to the sheet; but always cut them up the short way of the paper, or they will not bend properly. The paper for pinwheels and crackers is 30 inches by 20; the longest pipe that can be cut is 20 inches: as an experiment, try a pinwheel and cracker the other way of the paper. If the cracker is intended to be put into a jack-in-the-box, shell, or rocket head, push a bradawl up the mouth, insert a bit of quickmatch, to project a quarter of an inch, and secure it with a little wetted meal, pressed in with a knife. The way of bending it is shown at fig. 69. It is not necessary that a cracker should contain very much powder; the loudness of the bang depends more upon the thickness of the paper than upon the quantity of the powder. So with maroons; more string and less powder, are better than more powder and less string. A very good paper for crackers is 30-lb. royal cartridge; cut the strips 3 inches broad; 16-lb. double-crown may also be used, the strips 5 inches broad. Crackers for mines may be made with brown paper.

To make a cracker with 30 or 40 bangs, it is necessary to join 3 or 4 pipes together, before putting in the grain powder: the mode of making a joining will be understood by attending to the following directions. Cut a piece of paper 3 inches square, and lay it straight before you. At one inch from the right top corner, make a mark; at one inch from the left bottom corner, make a mark; draw an oblique straight line from one to the other, and cut along it with a pair of scissors. Without disturbing the relative position of the two pieces, draw the right piece a few inches towards the right; paste the farther edges, as usual; lay the wire close along the near edge of the right piece, and roll it up; it will have an external spiral; roll up the left piece in the same manner; it will have an internal spiral; when both are dry, paste the external spiral, and screw it, as it were, into the other piece. If managed carefully, and brought up till the edge of the paper forms a straight line, the joining will be as firm as if the paper had not been cut, presenting only a spiral edge, going once round the pipe.

Leader Pipes.

These are for piping quickmatch; they are rolled exactly like pinwheel pipes, on wires of different thickness, to suit the size of the match. They must be large enough to admit the match easily, without much pushing, which would break it; otherwise, the more nearly they fit the match, the more rapidly it blows through. The fire is conveyed from the tail of one case to the mouth of another, by a short piece of pipe, d, fig. 80, with the match projecting at both ends; when the fire has to be conveyed to two other cases at once, a fresh bit of piping is slipped on, and the match left exposed, as at a; or a piece is cut out of the side, as at b; each end should be bent into a hook, as at c, to prevent its slipping back. Wheel cases are to be enveloped by rolling a piece of double-crown twice or thrice round them, two inches longer than the case, as fig. 82, shown by the dotted lines. Suppose you have six wheel cases ready charged; lay six pieces of double-crown flat before you, and work them back with the thumbnail, as before described; paste the 6 edges; lay the 6 wheel cases in a row, side by side, and draw the paste brush across the middle, as if pasting your knuckles; then roll one in each envelope, so that it projects an inch at each end. Put the leader pipe in, and tie it with a piece of carpet thread, thin twine, or waxed yellow flax or hemp, as at fig. 83. To join one pipe to another, to lengthen it; suppose you have a piece of match 40 inches, long, and two pipes of 20 inches each. Slip both the pipes on the match, so that they touch in the middle; take the end of one, so touching, gather it round the match; pinch the end of the pipe tapering, push it into the other, and bind a piece of pasted paper round, to secure the joint. In cutting a matched pipe straight across, of course the match inside gets cut flush; do not leave it so, but, with a pair of pointed scissors, cut away a quarter of an inch of the pipe all round, and bend the exposed match, as at c, fig. 80, before putting it into the envelope of the case. Besides a strong pair of pointed scissors, a small pair, 3 inches long, with rounded ends, to carry constantly in the waistcoat pocket, will be found convenient for many purposes.

MAROONS.

Take 3 inches of a ⅝ or ⅜ rocket case. Fix, in one end, a cork, half an inch long; put in 2 inches of F grain powder; on this, another cork. Wind string tightly round it, lengthwise, 6 or 8 folds, side by side; bend it to a right angle, and wind 6 or 8 more folds; and so on, till covered; then wind crosswise; and again, at right angles, as in fig. 74. Dip it into melted glue, and put by to dry. When dry, make a bradawl hole through one part, to reach the powder. Make sure of this, by inverting it, and letting a little drop out. Insert a short port-fire, having a piece of match at the bottom, and touch-papered at the top.

Another Method.

Take a rocket case, 5 inches long, and unchoked. Put in one solid inch of plaster of paris. When this is dry, pour in 3 inches of F grain-powder; on this put another solid inch of plaster of paris. When dry, wind string tightly round it, up and down the cylindrical part, not the ends, till of two thicknesses. Paint the string over with melted glue. Make a bradawl hole in the middle of one side; fasten to it a squib, without a bang, having a piece of quickmatch at the end, to enter the hole. Cover the joining with two layers of pasted paper. On the other side, below, fasten a short piece of deal, like a rocket stick, with a point, to put into the ground, fig. 72. After lighting the touchpaper, remove to a distance; as bits of string are likely to get driven into the face, on the explosion.

To fire a salute with maroons, at regular intervals of time. Charge a port-fire, and saw it into inch lengths; envelope each in a piece of double-crown, 3 inches broad, and long enough to go thrice round the port-fire. Hang the maroons to hooks, or otherwise suspend them, on a frame, a foot distance from each other, as a, b, c, fig. 75. Underneath them fasten, with binding screws, or tie to nails, the port-fires x, y. Connect the port-fires with one another, and with the maroons, by leaders, in the usual way. On lighting at w, the first maroon explodes, and the first port-fire catches; this, having burnt, lights the second maroon and the second port-fire; and the port-fires being of the same length, the intervals of time between the explosions of the maroons will be the same.

GOLD AND SILVER RAINS.

These are little cases, $2\frac{1}{2}$ inches long, rolled on a $\frac{1}{4}$-inch former, and filled with the funnel and wire. They may be primed like fig. 29 or 30, or like squibs. Put them, mouth downwards, into rocket heads.

PEACOCK'S PLUMES.

These are a combination of rain and star. Roll them like pill-box cases, on a ⅜ inch former, about 1¾ inches long; charge one end, ⅜ of an inch deep, with coloured fire, driven in dry; fill up with gold, or silver rain, with a film of shell fuse at the top, to bind. Cut a bit of match, 2¼ inches long, lay it outside the case, so that it projects ⅜ of an inch at one end; envelope it in a piece of double-crown, fig. 90; tuck in the paper, to press the match, at one end, on the colour; twist the other to a point. Both ends thus light at once; and the rain appears, like a coloured star, with a tail.

To prime a case with match laid flat on the mouth. Take a piece of thread, or fine string, and fold, or bend it, in the middle, as at fig. 76. Tie a knot near the bent end, as at a, fig. 77. Bring the knot, a, up to the side of the case, as at fig. 78; pass the loose ends round, and tie in a knot, at b. Lay the bits of match flat across, as at e, fig. 79; bring the threads together, and tie them at c. Or bend tape match across, and tie, as at d.

To get a fine thread through a long pipe, or the hem of a bag. Take a piece of copper wire, and bend it round at one end, as z, fig. 88. Pass the end, z, forward, and push it through to the other end of the bag, &c., then bend it to the form of x, fig. 89; pass the string through the loop x, and pull the wire back.

SAUCISSONS.

These are a large kind of serpent, charged on a nipple, like a wheel case, with solid drift and mallet. They may be ⅝ or ⅞ or larger; about 3½ inches long. Drive in brilliant fire, or gerbe, fig. 84, 1½ inch; fill up to within ½ an inch of the top with F grain powder; and plug the end with plaster of paris, or a bit of wood, fastened with a tack or two. Press a piece of touch-paper, or double-crown, into the shape of a deep pill-box; fill it with F grain powder; fit it to the mouth of the saucisson; tie round the choke; brush, with meal paste, the outside, at bottom, and dip into dry meal. These saucissons are to be fired in a volley. Procure, say 2 dozen, iron tubes, a, b, c, &c., each a foot long; fit them with a wooden bottom, fig. 92, having a tenon, t, an inch long; let it be fastened with a screw on each side. Bore a hole through, to make a communication with the mortar formed by the tube. Take a board, an inch thick, of suitable length and breadth; bore in it 2 dozen holes, of a size to fit the tenons; glue these in, so that the tubes, or mortars, stand upright, in rows, side by side, like the pieces on a chess-board. Invert it. Nail a rim all round, so as to make a box, 2 or 3 inches deep. Cut a groove from hole to hole of the tenons; connect all the holes with naked match, also push a bit of match up all the holes in the tenons; now fill the box with sawdust, and nail a board on, to serve for a bottom, and to keep the sawdust in. Invert it; and put a saucisson, mouth downwards, into each mortar. Fig. 84 represents a single saucisson; w, w, w, fig. 85, saucissons in the mortars. On firing the match at s, it is evident the cases will be driven out rapidly, one after the other. The sawdust prevents the flash igniting the whole at once.

PEARL STREAMERS.

Paste brown paper all over, and roll up a case, of four or five thicknesses, on an inch and a quarter, or an inch and a half former, like a rocket or other case; when dry, cut it in the lathe (see fig. 27) into inch lengths—inch-deep bottomless pill-boxes. Set one on a foot, fig. 9, to enter about $\frac{1}{8}$ of an inch; mallet in the pearl streamer fuse, till nearly full, then a little meal powder; remove it from the foot, and press in flat, with a knife, a little plaster of paris, to form a bottom. They will have the appearance of bungs; fire them in volleys, like saucissons, from suitable-sized mortars. Primed end downwards, of course. Match may be tied on, as in figs. 78 and 79.

BLUE LIGHTS & STAR CANDLES, OR STAR LIGHTS.

These are little cases, charged with the funnel and wire; the latter are filled with spur fire.

PRINCE OF WALES'S FEATHERS.

These are simply pinwheel pipes, usually of coloured double-crown, charged with pinwheel fuse, and not wound on a block, but kept straight.

LANCES.

These are little cases charged with white or coloured star composition. They should be of white or coloured double-crown paper, rolled dry on the squib former, and secured at the edge with paste in the usual way. They may be from 2½ to 4 inches long, as may be required; press in one end to make a bottom. To do this, bore a hole through a piece of cork or a small bung, and through it push a piece of brass wire or stair-rod, fig. 81, of a suitable diameter. It can then be set to any distance, like a cutting gauge. If the lance case is 2½ inches long, set the bung at rather more than 2¼ inches from the end. Put it up the case, and, holding it with the left hand, with the right, with a piece of wire push in the end to make a flat bottom. Charge with the squib funnel and wire; and prime the mouth with very slightly damped meal. Lances are used for forming letters and designs, similar to gas jets. Frames for letters and devices are made with pieces of thin wood and cane, or hooping, for straight lines and curves. A number of wires, 3 inches apart, are driven into the frames so as to stand forward, at right angles, and the lances are fixed on them. The following is the way of proceeding. Procure some inch French nails, or inch rivets; the former at the ironmonger's; the latter at the grindery shops. Drive them in, at the proper distances; then, with a pair of cutting pliers or nippers, cut off their heads. Drive a piece of wire, the same thickness as the nails, into a bradawl handle, and leave it projecting ⅜ of an inch; file this triangular, or three-sided, and pointed. Push this triangular bradawl up the bottom of the lance, then fix the lance on the wire destined to receive it; it will be more secure if the wire is touched with a dip of glue before pressing the lance on. Having completed the letter or device, proceed to leader it. For this purpose have a supply of ½-inch rivets; they may be purchased at the grindery shops for about 5*d.* per lb. Take a length of leadered or piped quickmatch; lay one end of it on the mouth of a lance; push the triangular bradawl through the match and down into the lance; turn the bradawl round, which will insure the breaking of the match and of the priming; withdraw the bradawl, and push in a rivet; and so proceed. Take a strip of flannel 3 inches broad, roll it into the shape of a cork, and secure it from untwisting with a bit of string. Dip this into a solution of gum arabic or thick dextrine, and rub it over a sheet of double-crown. When dry, cut the sheet into pieces about half an inch broad and an inch and a half long, something like postage stamps. Take one of these, damp it like a postage stamp, and press it over the joining, and smooth it round the lance. Or a strip of paper may be pasted, and pressed round. The match will thus be nailed, as it were, to the lances; and prevented from slipping off by the gummed or pasted strip of paper. A common bradawl will not answer the purpose so well as a triangular one, as the sharp edges of the latter break the match and priming, and insure the ignition.

To Form a Device, Or Design.

Take a sheet of paper, and draw upon it a representation of whatever is intended, as a temple, a mosque, a ship, a horse, George and the dragon, &c.; then cross the design with lines, at regular distances, and at right angles, so as to cover it with squares, as fig. 86. It is now requisite to have a floor, of a considerable size; but not necessarily so large as the design intended to be fired, as a part can be done at a time; this floor must be divided into large squares, and the device from the small pattern, fig. 86, transferred to it, and the outlines chalked on the floor, fig. 87. After this a number of frames are to be made, of deal, or other wood, like square lattice-work, as fig. 91. A frame is then to be laid on the floor, so as to cover *a portion* of the design; and the French nails, or inch rivets, before mentioned, driven in, at distances of 3 inches, to receive the lances. Every frame must be numbered, and a copy kept in miniature, so that they may be correctly fitted together, to form the figure, without delay or error. To distribute the fire rapidly, over an extensive piece, it is necessary, at points, to make one leader light several, simultaneously. Suppose 10 are to be lighted at once: bring the 10 ends together and tie them; envelope them with a piece of double-crown, projecting a couple of inches; in the void put a scoop of meal; bring in the single match that is to light them, and tie as usual. Paste a piece of paper over all, to make secure.

To Preserve Steel Filings, or Cast-Iron Borings.

Put 1 lb. into a frying-pan, or iron ladle, with 3 ounces of marine glue; set it over the fire; and, as the glue melts, stir it about till thoroughly incorporated with the filings, or borings. When cold, bottle them, and cork. Marine glue may be obtained at Pattrick and Sons, 51, High Street, Whitechapel. It is 1*s.* per lb., which is the smallest quantity sold. Marine glue is made by putting pieces of india-rubber into mineral naphtha; the caoutchouc swells up; it is then to be triturated into a uniform mass, mixed with shellac, and melted. Wax solution, or stearine solution, rubbed up in a mortar, with steel filings, or cast-iron borings, also preserves them. The nitre of the gunpowder, however, attacks them, and ultimately rusts them; so that fireworks are never so brilliant as when recently charged.

SHELLS.

Shells are hollow paper globes, fired vertically, from mortars, or iron tubes. They are made of various sizes, from 3 inches in diameter to 16 inches. To make a 3-inch shell. Turn a wooden ball, 3 inches diameter; and, round the middle, that is, the equatorial circumference, cut a V groove, or triangular channel, deep enough to receive a piece of raw or naked match. Remove it from the lathe, and cut it into two halves at right angles to the groove, that is, round a meridional circumference. Construct a deal box, 4 inches square, 2 inches deep. Place one of the half globes, flat surface downwards, on the middle of the bottom of the box, and secure it with screws from underneath. Brush it, and the inside of the box, all over, with sweet oil, with a camel's-hair pencil. Put some water into a basin; sprinkle into it as much plaster of paris as judged necessary; about 4 tablespoonfuls; pour off the water which floats above; stir up the plaster till homogeneous; pour it into the box; and, with a sash-tool held upright, beat the plaster in with the points of the bristles. Leave it to set.

Instead of having a wooden ball turned, a hemispherical concavity may be made by pressing, half way, into sand, one of the painted india-rubber balls sold at the toyshops; and pouring plaster over it. Or, a basin, an inch diameter larger than the intended shell, can have the plaster mixed up in it, till about three-parts full; and then the bottom of an oil flask can be pressed into it. A narrow strip of blue paper should be previously pasted round the oil flask, at the proper height, as a guide to know the proper depth to which it may be pressed. The plaster, when partly dry, must be neatly trimmed; and may be left, permanently, in the basin. Or, a stiff paper cylinder, or a tin cylinder, may be made, an inch larger in diameter than the diameter of the intended shell: put the oil flask into this, neck downwards, and pour in dry sand, till only the hemispherical bottom of the flask is left exposed; level the sand; oil the flask; and pour in plaster, as before. Be careful that the mould is not less than half-an-inch thick in any part. Or, one or two, or more halves of the zinc, or copper globes, used for ball-taps, may be obtained of the plumber, and used for moulds, without further preparation.

To Make the Shells.

Procure two kinds of paper; one imperial brown; the other sugar paper, printed paper, paper hangings, or any paper of a different colour from brown. The shell is supposed to be 3 inches; half as much again is $4\frac{1}{2}$; add to this $1\frac{1}{2}$ for a flange, gives 6. Cut out a circular piece of the brown paper, 6 inches in diameter. Snip it all round with the scissors, in slits, reaching from the circumference, half way towards the centre; soak it in water, and lay it on a towel to drain. Have a piece of sponge, about the size of an orange; soak it also, and wring it. Place the piece of brown paper in the plaster concavity, and press it in neatly with the sponge, in all directions; it will fill up the mould, and overhang all round; press the overhanging part flat, so that it forms a flange. Cut a piece of another coloured paper, into a strip, about 2 inches broad, and paste it well; then cut it across, backward and forward, like the letter W, which will reduce it to V-shaped triangles. Take these up, one by one; lay them in the brown paper, pasted side downwards, and press them neatly in with the sponge, making each slightly overlap the other. Let these also overhang to thicken the flange. This being done, proceed with a layer of brown; and so on, alternately, till it gauges about $2\frac{3}{4}$ inches across. The shell will, then, be about $\frac{1}{8}$ of an inch thick. Remove it, and proceed with others. The different-coloured papers enable the eye to detect, in a moment, whether any part is left uncovered. When dry, cut off the flange; make the edge straight: rub it on a sheet of glass paper, spread flat on a board. With a $\frac{5}{16}$ inch punch, cut a hole in the middle of one hemisphere; to it, glue another hemisphere: and cover with two more layers of paper, or pieces of calico. The shell may, then, when dry, be filled with any kind of stars, or rains, that can be got through the hole. Along with the stars put $\frac{1}{16}$ of their weight of meal powder for a bursting charge: that is, if the stars weigh 4 ounces, as they probably will, put 4 drams of mealpowder. Charge a roman candle case, $\frac{7}{8}$ inch internal, $4\frac{1}{2}/8$ external diameter, with shell fuse (see fig. 59); saw it into inch lengths. Cut a piece of calico, 2 inches broad, and long enough to go twice round the fuse: paste the calico all over, and roll the fuse in it, so that the envelope is flush at one end, and overhangs one inch at the other. Glue this in the hole of the shell, the flush end, a, fig. 93, of course, inward: the enveloped end, b, being to receive the match, $\frac{3}{4}$ of the fuse may be pushed in, $\frac{1}{4}$ of an inch left to project. With a printer's bodkin, or a stiletto, make a hole through both sides of the envelope, as at c and d; these holes are to be in a line with the groove; put through them a piece of raw match, in such a manner that it shall lie across the mouth of the fuse, and go, in the groove, round the shell underneath, and reach to e and f. Paste strips of paper over the match, the same as with tourbillions and saxons; or, the part of the match lying in the groove, may be in a leader pipe. Weigh the shell, and take about ⅕ of its weight of coarse grain powder, for a blowing charge. Make a cone of two or three thicknesses of paper; put the blowing charge into it; stick the cone on the bottom of the shell, and set by to dry. In the calico mouth of the fuse, tie a long piece of leadered match, and paste a strip of paper round, to make secure. If the mould has been made with the indentation of an oil flask, or with the globe of a ball-tap, it will, of course, not have a groove to receive the match; but this is of little consequence; the match can be drawn round outside, and covered, so as to appear as it does on tourbillions and saxons, like a vein on the back of the hand, when the fingers are held downwards.

Formerly mortars were made of sheet iron, riveted and bound round with cord, which latter would not prevent them cracking, if they were not thick enough: they are now made on an improved principle: the iron is rolled, by powerful machinery, of three thicknesses, exactly like a squib case; it is, then, made white-hot, and the three are welded together, with a steam hammer. Large mortars, also, have an iron bottom, or breech fixed in them, and are farther strengthened with a couple of rings, put on hot, and shrunk by cold, like tires on wheels; a third ring is put over the other two, as in the Armstrong guns. The mortar is placed in a hole, dug in the ground, a few inches left standing out; the earth is shovelled in, and driven down firm; a penthouse lid, to keep out rain, dirt, and insects, renders it complete. Amateurs require nothing of this kind. A small mortar may be a tube, open at both ends, and fitted with a wooden bottom, to which it is to be firmly screwed. Fig. 94 represents such mortar: it may be 4 diameters high; and the foot should have a conical hole turned in it to receive the cone fastened to the shell. The match is lit at t; but this may have a long bit of touch-paper attached to it, if preferred.

Instead of making a plaster mould, to form the shells in, the shells may be made by covering a wooden sphere, with paper, on the outside; when dry, they may be cut round in the lathe (a cross mark, with a pencil, having previously been made, as a guide to bring the same parts together again); the wooden mould removed; the cut edges glued; and the shell fitted up, in the usual way. Clean oil flasks may be covered with six or eight thicknesses of paper: paste an inch, or two, round the neck; when dry, cut through the cover, near the spherical part; file a notch all round, and snap it off. I have made excellent shells this way; the chief objection against them is their limited size. Glass globes might be blown, of uniform size, in moulds, like bottles. Another ready way of making shells, is to cover the india-rubber air-balls, of the toyshops, almost as thin as soap-bubbles; when the cover is dry, a hole may be cut, for the fuse, with a penknife, and they are ready, at once, to receive the stars. Their shape is that of a prolate spheroid, fig. 106. After eight thicknesses, or more, of paper have been pasted on, measure, with a tape, round the equatorial circumference, b e d, which suppose 17 inches; add 1 to this=18 inches. Measure from the pole, a, down the meridian by e to the opposite pole c, suppose 11 inches. Cut a piece of double-crown, 18 by 11; fold it down the middle, to a double thickness of 9 by 11; fold-again to $4\frac{1}{2}$ by 11; again to $2\frac{1}{4}$ by 11; there will, now, be eight thicknesses. Pencil the shape fig. 107 upon the top, and cut through the whole. Paste the eight gores on, as in fig. 106; for ornament, half may be pink; half, green. If these air-balls could be blown in a spherical mould, of uniform size, they would obviate the gluing process, which is, at present, a tedious and necessary evil; they would, also, be much cheaper, as they could be supplied for about 4*d.* per dozen, and save the cost of a great deal of needless labour.

CYLINDRICAL SHELLS.

A sphere is, by no means, the best shape for a projectile; no one would think of making a roman candle star like a marble; the Minie bullets and the bolt-shot for our great guns are cylindrical, and far better fitted than globes for straight and rapid flight. Why not adopt the same shape for shells? I find cylindrical shells answer excellently. Have a former, for a small one, 2½ inches diameter, and about 6 inches long, and with a handle like fig. 25. Roll a case upon it, 5 inches long, till the outside gauges 2¾ inches, or a trifle more. Turn a wooden bottom, ¾ of an inch thick; half of it a tenon to fit the inside of the case, and half of it a flange, equal to the external diameter, a, fig. 95. Glue this firmly in, and farther secure it with 4 inch French nails; though, perhaps, this is not necessary. The top may be of the same shape; half-an-inch thickness will be sufficient. A hole, an inch in diameter, may be bored in it, with a centre-bit, to receive the fuse. This fuse may be a cotton reel, with one of the flanges sawed off, and the end filed slightly tapering, as fig. 105; the enveloping piece of calico, or glazed lining, can then be passed to the bottom, as indicated by the dotted lines; but a better way is to turn a piece of beech of the shape of fig. 96, with a collar, to prevent its being blown through, and a groove, by which the calico envelope can be tied. The length of the fuse may be about 1¼ inch; charge it by putting in very little at a time, and well mallet it. Pass a piece of naked match over the mouth of the fuse, and down the sides of the cylinder, as indicated by the dotted lines, fig. 95, having previously cut out a notch, at each corner, top and bottom, e, e, e, e, to guide it in a straight line. Take a piece of double-crown, about 20 inches long, and 7 broad; paste it all over, and roll the cylinder in it, in such a way that it shall be flush at top, and project at the bottom; press it round smooth with a cloth, or sponge, till the leaders form a vein on each side. When dry, invert it; put in the blowing powder; press the ends over, neatly, to form a bottom; and cover the whole with a circular piece of pasted paper. A shell of this size will hold forty gold rains, and a score or more blue stars, which have a very pretty effect. The head and the fuse may be in one solid piece, if preferred, as fig. 104; if the hole is longer than 1¼ inch, only 1¼ inch of it must be filled with fuse; a piece or two of match may fill up the rest. The mortar for this shell should be 18 or 20 inches high, and 3 inches internal diameter. The shell, when fitted up, will probably weigh about 14 or 15 ounces. The blowing powder may be 2 or 2½ ounces of F grain, according to the fit in the mortar. The hole, in the wooden fuse, should be ⅜ of an inch diameter. See that the match is everywhere carefully covered. A shell of this size will be amply sufficient for amateur purposes. Perhaps it will be advisable to try a fuse an inch long the first time; it will be better that the shell should burst, while still ascending, than that it should pitch. One or two trials are, in all cases, necessary; but as these shells go a great height, they will bear a little longer fuse.

A cylinder holds half as much again as a shell of equal diameter; consequently, when the depth is twice the diameter, which is the best proportion, it holds three times as much as the spherical shell. If the latter be 3 inches diameter, it may be represented by the numbers $3 \times 3 \times 3 = 27$. If the cylinder be 3 inches diameter, and 6 inches deep, it will be as $3 \times 3 \times 6 \times 1½ = 81$. Practically, it will hold more, especially of rains, or serpents, as they pack better in a cylinder.

The spheroidal shell, like the spherical, is, also, ⅔ of its circumscribing cylinder.

Putting D, the diameter of a spherical; and d, that of a cylindrical shell, the length being always double the diameter; then $D^3 = 3\ d^3$. Required the diameter of a cylindrical shell, that shall be of equal capacity with a spherical shell, of 12 inches diameter.

$12^3 = 1728$; $1728 \div 3 = 576 = d^3$, and

$\sqrt[3]{576} = 8 \cdot 3 = d.$

so the cylindrical shell will be 8³⁄₁₀ inches diameter, and 16⅗ inches long.

The same computation is more readily effected by multiplying the diameter by $\sqrt[3]{(1/3)}$;

$\sqrt[3]{(1/3)} = \sqrt[3]{(9/27)} = \frac{1}{3}\sqrt[3]{9} = \cdot 693$.

Then $\cdot 693 \times 12 = 8 \cdot 3$, as before.

Shells, for war purposes, are both spherical and cylindrical; the latter are, sometimes, built up in pieces, which dovetail into each other; they are, then, set in a cylinder, and melted lead is poured round to bind the pieces together. For lighting up the country, to discover the movements of the enemy on a dark night, shells are fired, containing strong calico parachutes, carrying blue lights. Magnesium lights are, of course, more effective.

The bottoms and tops of the cylindrical shells, previously described, may be turned out of a plank of wood: elm, ash, common mahogany, or good yellow deal; and, as many amateurs who possess a lathe, know of no contrivance for holding flat pieces of board without making a hole through them, the following plan may be adopted. On the screw-chuck, fig. 102, screw a piece of deal board, previously sawn nearly circular, and as large as the lathe will take: let it be of a thickness to entirely hide the point of the screw; turn it to a circle; and over the face of it, with a blacklead pencil, while it is revolving, make a number of concentric circles, as in fig. 103. Also, cross it with two straight lines, passing through the centre, at right angles to each other. Cut the wood, intended for the bottom of the shell, into a square; make a hole at each corner, and with 4 screws, or 4 French nails, screw or nail it on the face: the concentric circles, and the two diametrical lines crossing them, will act as guides to centre it. I have thus been able to cut into a circle, in the lathe, so thin a substance as a piece of writing paper. It may be held on with 4 tin-tacks, or a touch of gum.

A nest of 6 or 9 little shells, each containing a separate colour, rains, serpents, crackers, &c., may be enclosed in a large shell: the fuse of these may be a very short piece of a squib-case, with the string wound 7 or 8 times round the choke, to form a flange, to prevent its blowing through.

Cylindrical shells should, of course, be filled before the head is glued in; this can have the fuse previously fixed in, and ready.

ASTEROID ROCKETS.

These differ from other rockets only in the head, which contains a parachute for floating a coloured case, attached to the strings. The parachute is best formed of silk, or alpaca; it may also be made of black glazed lining, or of tissue paper, carefully rubbed, till the stiffness is removed. The colour should be dark, that it may be invisible. If of silk, or alpaca, the parachute may be 20 inches square, with 4 strings attached to the corners; or it may be made of 6 triangular pieces, sewed together, like an umbrella. If of tissue paper, the paper may be $\frac{5}{8}$ of a circle, or $\frac{5}{8}$ of an octagon, as fig. 97. The tissue paper must be strengthened with a piece of crochet cotton, running round the edge, as shown by the dotted lines, fig. 101. It is simply laid on, and the paper snipped at each corner, with the scissors, pasted, and turned back. The radius of the circle may be 15 inches.

For the lance, or colour, roll the case of 3 thicknesses of writing paper, on a $\frac{5}{8}$ former: it may be $2\frac{1}{2}$ or 3 inches long, and pasted all over. Turn a little wooden pulley, a, fig. 98, of a diameter to nearly fit the cylindrical head of the rocket; a hole in the centre of this pulley receives the colour; round the pulley, in the groove, lay a piece of chenille, fig. 100; twist the wires of the chenille 2 or 3 times together, and cut off flush. To charge the lance, set it on a flat surface, and drive in a film of meal powder, or shell fuse; then $\frac{1}{3}$ of an inch of crimson star composition; $\frac{1}{3}$ of an inch of green, &c., till nearly full; stop the end with plaster of paris, pressed in flat, with a knife. Paste a strip of paper round the mouth, so as to overhang not more than $\frac{1}{8}$ of an inch; lay in two short pieces of match, one across the other, and tuck in, or press down, the pasted overhanging edges; this will keep the match from dropping out. Glue the pulley on, about midway of the case; envelope the end b with a bit of glazed lining. Take a piece of string, fig. 99; bring the ends together, and tie in a knot. Insert the knot in the envelope, and tie round above it, as at fig. 100, leaving the loop out, to which to attach the strings of the parachute. Let the strings be 2 feet long: fasten them to the loop; gather them together; wind them round the colour, above the pulley; fold the parachute neatly, and bring the edges down over the strings, nearly to reach the pulley: this will keep the parachute a little open, and help it to expand. Both the tissue paper and the glazed lining should be well rubbed, to get the stiffness out of them. To ascertain how a parachute will act, the learner can take the colour case, and stop one end with plaster, fill up with sand, and stop the other end with plaster. Fire it in the day time. Silk, alpaca, and glazed lining parachutes I can vouch for; tissue paper I have never tried, but am told it answers if it has not remained folded so long as to get set.

Between the top of the rocket and the mouth of the colour is to be a little pad of grain powder. Make a little square or circular paper bag with double-crown, and having put in the grain, and made the joining secure, smear both sides in the usual way, and sift dry meal over. If the rocket is to be fired at home, the parachute may be put into a long cylindrical paper bag, and left quite open at the top; but if intended to be carried to a distance, a thin bung may be put in at top, or a piece of turned wood: or a thin piece of paper may be pasted on it. In this case, be careful that it does not get stuck to the parachute.

Instead of the cylindrical head and the chenilled pulley, the head, fig. 39, may be used; and the protecting power from the fire to the strings supplied by putting a quantity of bran or sawdust. The cone must be held on, with not more than two thicknesses of double-crown.

COMPOUND FIREWORKS.

Fig. 108. A triangular: 3 wheel cases and a case colour, at a. These cases are tied either to a hexagonal piece of deal board, or to three spokes, radiating from a nave. The colour is sometimes tied to a nail, driven in to receive it; or, it may be fixed on to a little peg. The peg is shown, by the side, at b. It is turned with a tenon, c; this tenon is glued in a hole, bored in the spoke; the part b may be half an inch in length; through it bore a small hole; charge the case colour on a foot, that enters the case half-an-inch; this void fits on to the peg b; pierce the case, through the hole in the peg, with a bradawl; push a bit of binding wire through, and twist. This is a far better way than tying it to a nail, as it cannot drop off. If the colour gets blown off, as it often does when tied, the piece is half spoilt. The colour may either stand at right angles, so as to face the spectator; or it may be fixed so as to lie in a plane with the wheel cases, very slightly sloping upwards; the mouth of the colour should point the same way as the mouth of the cases; if it were placed in the opposite direction, it would meet the current of air, be blown back upon itself, and burn the case rapidly. With respect to the wheel cases, it is obvious that the mouths must all point in the same direction; the slightest consideration would suggest this; it is necessary, therefore, to have some certain rule of proceeding, so as to avoid mistakes. Now supposing, for the mere sake of illustration, if you had a case in your hand, it were to take fire, you would naturally wish the fire to be directed *from* you. Let this, then, be the guide. You are about to tie the cases on a wheel. Sit, to do so; take the wheel between your knees: place a case upon it, with the choke end *from* you; tie, near the choke, also near the end; turn the wheel, place another case upon it, choke end *from* you; and so forth. Without this certain way of proceeding, you would be very embarrassed with the 12 cases on fig. 112; but, by attending to it, not the slightest difficulty will be experienced. Some of the cases will point obliquely upwards; some downwards; but they will all point *from* you. In the end of the last case, as at x, fig. 108, put a little dry clay, to prevent a stray spark igniting it, and tie the envelope; or, tuck in the envelope, like as with an ounce of tobacco. Avoid every source of failure. Even the knots of the strings, with which the cases are tied on, are apt to come undone; they should either be touched with a dab of glue, or have a piece of paper pasted over them.

Fig. 109. A double triangular: 3 cases tied to the spokes; 3 to the previous 3. Double triangular frames are also made, with 6 spokes, on a long nave, 3 behind 3.

Fig. 110. A vertical wheel, illuminated. This is a wheel, with spokes, and a rim, or felly. The wooden hoops of the toyshops will furnish the latter. The illumination, as it is called, is made by lances, nailed with ½ inch Flemish tacks to the nave. The lances should be of different colours, but they ought, as nearly as possible, to keep time with each other. To effect this, it is necessary previously to adjust them; thus. Charge a number of cases, exactly two inches long, with different colours; use the same scoop to all, and give each the same number of blows, that they may be rammed as uniformly as possible. Fasten a string to the end of a leaden bullet, and tie a loop at the other end of the string. Let the length from the top of the bullet to the top of the loop be 39 inches. Suspend this from something, and set it swinging. Light the lances, one after another, and count the number of oscillations each endures. Keep a list of these, and write against them 10, 12, &c., or 5 per inch, 6 per inch. The bullet will indicate seconds, with the length of string recommended; and it matters not whether you give it a start of 6 inches, or 12 inches; for, if it goes twice as far, it goes twice as fast, so that the swings isochronise; the inestimable discovery of Galileo, which led to the invention of the pendulum. In order that the lances may be nailed on, they should be charged upside down, and left with ½ an inch vacancy; they can then be pinched flat, to receive the tack. Scrape out a little from the other end, and prime with very slightly damped meal.

Fig. 111. A rainbow wheel. This is a vertical wheel, generally with 3 colours, as drawn; the tail of the second, or mouth of the third, lights a; the mouth of the fourth, b and c; but any arrangement may be made. Place the colours, red, green, blue, at different distances from the centre, so as to form rings, equidistant, when burning. Suppose the spoke 12 inches long; place the colours at 3, 6, and 9 inches from the centre. It receives its name from its resemblance to the rainbow.

Fig. 112. A caprice, or furilona, according to the number of the cases. A caprice, from the capricious manner in which it turns, up, down, and round about, now this way, now that. A furilona, possibly, from the fury with which it plays, when 4 cases are burning together, at the end: though some call it a fruiloni, said to be from the name of its inventor. A furilona and caprice wheel are much the same; the former, generally, has fewer cases on it than the latter. A coloured gerbe, placed on the top, is very effective; or, it may have a mine, or jack-in-the-box. The cases are to be placed so that some of them play horizontally, some obliquely upwards, some obliquely downwards; the spokes, which are concave at the end, are glued in, so as to determine the slope of the cases. If there are 10 cases, they may be fired thus, h, u, d, hud, hudp; that is, 1 horizontal, 1 up, 1 down; 3 at once, horizontal, up, down; 4 at once, horizontal, up, down, perpendicular. If 13 cases, thus: h, u, d, hu, hd, ud, hudp. While tying the cases on the frame, it should be on a short wire, held perpendicularly in a vice, or block of wood; properly, the wire should be tapering, so that the frame should bite, when dropped on; the tapering will allow it to be turned round, easily, by slightly lifting it. Let the leaders be drawn straight, and not left dangling in curves, nor crossing each other. There must be *enough* match, but there ought to be no more. Look well to the mouths and tails of the cases; it is best to put a piece of pasted paper over each, for as the piece dashes round with great violence, if a stray spark falls on any composition filtering out, the whole is spoilt.

Fig. 113. A horizontal wheel, with mine and roman candles; the cases on the wheel are to be tied so that some play horizontally; some, obliquely upwards; some, obliquely downwards. To make a case play thus, tie it on the wheel, across the middle of the case; this being done, turn the mouth upwards or downwards, and tie it again, making the strings cross the previous ones, so as to form a letter X.

Fig. 114. A turning sun: two concentric hoops nailed to spokes, or a frame; the cases lying on the hoops, slope; 3 or 4 may light at once; the spokes carry triangular, or vertical wheels; at the centre is a double triangular, or larger vertical.

Fig. 115. A rayonant star piece: a wheel with six spokes; at the end of each spoke, two fixed cases, forming a V, the alternate spokes carrying saxons; at the centre, a double triangular, half-way between the triangles and the saxons, six five-pointed stars. A very beautiful piece.

Fig. 116. A chequer-piece: a true-lover's-knot in the middle, 16 fixed cases, 4 on each side the true-lover's-knot; and 4 saxons at the extremities. The fires cross, and chequer into squares.

Fig. 117. A scroll wheel: six or more cases on the wheel, to play in pairs; lances arranged on cane or hooping to form a scroll as indicated.

Fig. 118. A pyramidical piece: a scroll wheel in the middle; five horizontal wheels, or triangulars, at intervals, as represented by the rings; brilliant fixed cases playing obliquely upwards: at the bottom may be a row of cases playing downwards; these form what is called a cascade. Gerbes make the most effective cascades, but they require to be placed at a great height from the ground, if containing iron: the coke grains will be found suited for 8 or 10 feet.

Fig. 119. A spiral wheel: six cases on a horizontal wheel; lances arranged in a spiral, on cane, or hooping.

Fig. 120. A true-lover's-knot: six $6/8$ wheel cases, playing in pairs; three saxons, one carrying a blue; one, a green; one, a crimson colour. Light at a; this leader blows across, and lights the opposite starting case. The tail of this case lights the saxons: the ends of the saxons at c, c, c, before enveloping them, are to be smeared with meal; the end, b, is also to be smeared with wetted meal, to insure the ignition of the leader. This is a most beautiful piece: the colours, on the saxons, form loops, and represent, in a slight degree, the compound motions of the moon and planets, with regard to the earth. The centres of the saxons are carried round in a circle, like the earth in her orbit; the colours on the saxons revolve round the flying centres, like the moon round the earth. The wheel must not be less than 3 feet diameter.

Fig. 121. A revolving globe. This is, also, a most beautiful piece. The bottom is a horizontal wheel, carrying a strong half hoop, a b c; a skeleton globe, formed with hoops, is suspended in this. This globe is driven by cases placed upon it, round a hoop, crossing the other hoops, at right angles, like the equator, at right angles to the meridians. The meridional hoops are covered with lances, white or coloured. The globe revolves vertically, while the wheel below turns it horizontally; the compound motion produces a peculiar oblique tumbling convolution, exceedingly perplexing to spectators, ignorant of its construction. Instead of a globe, the top piece may be a revolving cylinder.

Fig. 122. A mine. This is a cylindrical case, containing serpents. The bottom of the mine should be a circular piece of wood, glued in. On it, place a circular bag, containing F grain powder. The bag is made with two circular pieces of paper, one half-an-inch diameter larger than the other; lay the small one on the top of the large one: paste, or gum, the exposed rim of the bottom piece, and bend it up, and press it down on the small or top piece, all round, leaving a part through which to put the powder; when dry, put in the powder, and stop up the hole. Put it into the mine, smear it with the brush, dipped into meal paste, in the usual way; and, with a pepper-box, shake in a little dry meal. Take a fixed case, charged; envelop it, so that the paper projects about an inch at the bottom: take a piece of squib-case, the same length as the serpents; put through it a piece of match, long enough to protrude at the top, half-an-inch, and to bend over, to form a hook: tie this in the envelope of the fixed case. Fill the mine with serpents, naked primed mouth downwards: with the scissors, or a pair of pliers, draw out the middle serpent; put in the matched squib-case; hold the fixed case upright, in the mine, and ram pieces of torn paper tight round it, to offer resistance, and cause the serpents to be blown higher. To adjust the blowing powder in the bag, use the following formula, I denoting the diameter in inches.

$I \times 2I$ = drams.

Required the quantity of powder for a mine $1\frac{1}{2}$ inch diameter.

$1\frac{1}{2} \times 3 = 4\frac{1}{2}$ drams = $\frac{1}{4}$ oz.

For $1\frac{3}{4}$ inches?—$1\frac{3}{4} \times 3\frac{1}{2} = \frac{7}{4} \times \frac{7}{2} = \frac{49}{8} = 6\frac{1}{8}$ drams = $\frac{3}{8}$ oz.

For 2 inches?—$2 \times 4 = 8$ drams = $\frac{1}{2}$ oz.

For $2\frac{1}{2}$ inches?—$2\frac{1}{2} \times 5 = 12\frac{1}{2}$ drams = $\frac{3}{4}$ oz.

For 3 inches?—$3 \times 6 = 18$ drams = $1\frac{1}{8}$ oz.

It is by no means necessary to be exact; but the formula will serve as a guide.

Fig. 123. A jack-in-the-box. This is a case formed on a square prism of wood. The paper is to be pasted all over, but as it cannot be rolled, it must be folded over, one side at a time, and rubbed smooth with a tooth-brush handle, or pressed with the fingers. The bottom may be formed by bending in, to a right angle, two opposite sides, and pressing them flat; then, upon them, the two remaining sides, like packets of cocoa, &c. Or a square piece of wood may be glued in for a bottom. The same blowing charge as for mines.

Fig. 124. A devil-among-the-tailors: a mixture of crackers and serpents; a roman candle in the middle, having its bottom stopped with shell fuse, instead of plaster of paris: 3, or 4, or any number of roman candles, at regular distances, round the outside: these are to be tied with string, and a strip of pasted paper fastened round.

Fig. 125. A line-rocket: two rockets tied to a piece of roman candle case, head to tail: that is, one rocket tied in one direction, the other, in the reverse. This simply runs up and down a line.

Fig. 126. A pigeon frame (of which fig. 127 is a cross section) may be made by taking a piece of deal, or alder, say 6 inches long, and $1\frac{1}{2}$ inch square: bore a hole through the length to receive the line upon which it is to run; plane off the corners, lengthwise, and channel them with a gouge, to form 4 semi-cylinders for the rockets to lie in: in the middle glue 4 spokes, each about 4 inches long; and round the spokes nail a wheel. A piece of cane makes good spokes, light and strong. Tie on the frame, lying in the channels, four $\frac{5}{8}$ rockets, two pointing one way; two, the other: the clay or plaster of paris, in these rockets, is not to have a hole through it, as usual, but to remain perfectly closed. On the wheel, tie four $\frac{5}{8}$ wheel cases, charged with wheel case fuse, No. 1 or No. 10. Leader it thus: touchpaper the mouth of one wheel case. From the tail of this first wheel case, carry two leaders; one to the next wheel case, and one to the rocket whose mouth is nearest: the tail of the second wheel case lights the third wheel case and second rocket; and so on. A long clothes line must be stretched tight for this to run along; it should start 3 or 4 yards, at least, from the post to which the line is tied: and the rope should be so long that the first rocket could not carry it to the end; otherwise the leaders are apt to snap and spoil the piece. Towards the end, as the piece gets lighter, it will run the whole length of the rope, without injury. It is necessary to indicate the first, or starting rocket, by some means; a piece of red paper may be pasted round the mouth; or, an inch or two of string may be tied to it; or the match may be bent and tied so as to project an inch or two longer than the others. On fixing it on the line, in the dark, you can then feel which is the starting rocket, and place its mouth *towards* the starting post, without hesitation. Of course the pigeon runs away in the opposite direction.

Fig. 128. A pigeon-house, made with lances; this may be fixed in a central position, and have 4 lines running from it: east, west, north, and south. As soon as the lances are well alight, start the pigeons. This is not necessary for amateurs. One pigeon is sufficient, without any pigeon-house.

The piece, fired at the Crystal Palace, termed a comet, is a combination of the pigeon, fig. 126, and the turning sun, fig. 114. Construct a pigeon frame, with two hoops, one at each end; to these hoops, attach brilliant cases, or gerbes, placed obliquely: point the mouths of all the rockets and gerbes in the same direction. The whole of the cases may be lit at once; or a short case may fire the rockets after the piece has partly descended the line, by its own weight. The line, at the Crystal Palace, is stretched from the top of the northern tower to the ground.

Fig. 129. A double guilloché, or windmill piece. This represents two windmills, turning in opposite directions; and imitates the engine turning, on the backs of watches.

Fig. 130. A five-pointed star. This has been already described.

It is desirable, sometimes, to convey the fire from a movable or rotating piece, to a fixed, or second rotating piece, which is effected thus. A leader, a, fig. 131, comes from the tail of the last case of the first rotating piece. This leader is tied, in one or more places, to one of the spokes of the wheel, and to the nave; and is left protruding at b. A tin box, or a bit of a mine case, d e f g, is attached to the fixed, or second movable piece in such a manner, as to surround, but stand clear of, the end of the match, b, as it revolves, with the wheel.

A hole is made at c, and the end of a leader, from the second piece, brought through it. The inside surface of the box, d e f g, is smeared with meal-paste. When the leader catches at a, it blows through to b; this lights the smeared surface of the box; the flash communicates the fire to c, and this carries it where desired.

Sometimes a piece is made to drop, which is thus effected. Suppose a cylindrical piece of iron, 6 inches long, 3 inches diameter, standing upright; on this a second cylinder, 12 inches long, 2 inches diameter; and, on this, a third cylinder, 2 inches long, and 1 inch diameter. If now a brass ring $1\frac{1}{4}$ or $1\frac{1}{2}$ inches diameter be put on the top, it will slip down, and rest on the 2-inch cylinder; if a ring $2\frac{1}{2}$ inches diameter be slipped over the top, it will fall, and rest on the bottom 3-inch cylinder. It is obvious that, if a horizontal wheel, having a hole in the nave, $2\frac{1}{2}$ inches diameter, be slipped on, it will drop, and rest on the bottom cylinder; and that a second wheel, with a $1\frac{1}{2}$ inch hole in the nave, will rest on the second cylinder. The two wheels, then, may be placed on, and held together with a piece of thickly leadered match; the top wheel plays for a while, and carries the second wheel with it; the tail of the first, or other case, lights the bottom wheel, and at the same time blows the match to pieces, and the wheel drops. Of course three or more wheels may drop, one after another. See the design, fig. 166.

Sometimes a horizontal motion is changed to a vertical, thus. Suppose an upright wooden post, 3 inches square. Saw off the top 18 inches, and fix it on again, with a hinge; drive a staple, on the other side, into the top piece; also one into the bottom piece; and connect the staples with a piece of string, to prevent the top piece falling. In the top drive a spindle, as usual, to receive a horizontal wheel. On the side of the fixed post, against the hinge, fasten an iron bracket, to stand forward, so that the top part, when the string is cut, can fall down, and rest on the bracket. Now slip over the spindle a brass tube, 3 inches long, and almost a fit; then put on it a horizontal wheel, carrying a smeared box, as above described, to convey the fire to a fixed piece; on the top of the spindle screw a nut. To the bottom post, tie a piece of shell fuse, and let the string, that prevents the top piece from falling, be drawn over the mouth of this, and secured. It is obvious that if, after the wheel has played awhile horizontally, the shell fuse is lit by a leader, the string will be burnt in two; the top part will fall on to the bracket; and the horizontal motion will be changed to a vertical one.

Roman candles are best fired in bouquets of three or more; connect them with leaders about 3 inches long; and set them in a block of wood, containing three or more holes, diverging right and left; or tie them to struts of wood, nailed together, like the supports for garden flowers.

Two vertical wheels may be made to run round the outside of a table, to imitate the motion of the machine for grinding drugs, crushing clay, &c., like two parallel grindstones going round in a circular trough. Conceive a circular table, 20 inches diameter. With a gimlet bore a hole in the centre, large enough to admit a screw-eye, with liberty to turn easily round; slip through it a stair-rod, 2 feet long; it will overhang 2 inches at each end, and may be swung round. At the distance of about 4 inches from each end of the rod, imagine a cotton reel, or a pulley; when the rod is moved round, the reels or pulleys will roll round in a circle, like the cylinder-crushers, above mentioned. Now, if a vertical wheel were fixed at each end of the rod, so as to hang outside the table, it is obvious that, upon firing it, the wheels would run round. Of course it must not be a screw-eye and a stair-rod; but anyone will understand, from this description, how to effect it.

These contrivances have fallen somewhat into disuse, since the introduction and variety of colours, but it is well to understand them.

To Fire a Girandole of 100 Rockets at once.

Suppose a cubical tea-chest. In the top, bore ten rows of holes, ten in a row, with a centre-bit; the same in the bottom, in such a way that the bottom holes fall perpendicularly under the top holes. Fasten the box upon four legs, one at each corner. Sift from a pepper-box a layer of meal powder over the top; put in the rockets, with their primed mouths, naked, to rest on the sifted meal. It is evident that, upon conveying fire to the meal, by a leader, the flash will ignite the whole of the rockets, at once. Of course it ought not to be a tea-chest, but a box constructed on purpose, with a penthouse lid, to fall over, and protect the rockets, till desired to be fired. It may be furnished with four legs, as in fig. 173. The shelf, fig. 172, fits in at a b c, fig. 173.

TOOLS.

Fig. 132. A port-fire holder. This is made of steel, somewhat after the manner of a pair of pliers. The ends a and b are fluted, or channelled (semi-cylindrical) to receive the port-fire: a spring, riveted at c, holds it tight; by pressing on d, the burnt case drops: the end e is pointed, to enter a long stick, bored to receive it, and strengthened with a brass ferrule.

As an instrument of this kind can be obtained only when made to order, the following contrivance will serve the purpose of most amateurs. First charge some little port-fires, 3 inches long and ¼ inch diameter, till within about 1¼ inch full: having filled a couple of dozen, or so, invert them, and knock out the dust, as with squibs; then encircle the whole, as if going to bang them; but pour in dry sand instead; empty a little sand from each, and stop the end with plaster of paris. Scrape out a little composition from the other end, and prime it with damped meal. Take two inches of a roman candle case, or a piece of turned wood, with a hole through it, fig. 133, two inches long. Have a long stick, fig. 134; cut the end, a, so that it will fit the hole in fig. 133, and enter half way up. At the other end fix a wire, z, two and a half inches long. By slipping fig. 133 on fig. 134 it is evident one inch vacancy is left to receive the sand end of the port-fire; when this is burnt out, it can be pushed out with the wire, z.

TO CONSTRUCT A STEELYARD SCALE.

The numerous formulæ for coloured fires, renders the use of very small scales excessively troublesome; and there is perpetual danger of losing the little weights for grains and pennyweights. By constructing a steelyard scale, the greatest facility is attained, one weight answering for all.

Take a piece of deal, fig. 135, about as thick as a venetian-blind lath; it may be 20 or 24 inches long; an inch broad, at the left end; and about ½ an inch broad in the main length. Make a hole at a, twice the diameter of a pin: the same at b. It is necessary to observe that the hole, a, must be on a level with the top of the long arm; if it should be a little higher, it would not be of any consequence; but it must, on no account, be lower. Have a piece of deal, fig. 136, about an inch square, and mortise a hole in it, as drawn. Fix 135 in 136 so as to appear like 137. A bit of a common pin, or needle, is to go through the mortise and through the hole, a, of 135. See that it plays easily. At s, fig. 137, insert a piece of wire, bent like the figure 8; and to it, tie the strings fastened to w, which may be a copper bowl, or a tin patty-pan. The mode of fastening the threads to the figure of 8 wire, is simply to slip them through, bend them down, and tie round with thread, as at z.

Take a piece of brass tube, an inch, or so, long, fig. 138, and with some lead, melted in the bowl of a pipe, fix a bit of wire in it, bent like a staple, or the capital letter U, as drawn. This is for a weight. Slip it on the arm of fig. 137; put a wooden pin in the hole y, to keep the weight from falling off. Slip the weight along, till it balances the patty-pan w: suppose this point is at m. This will be the starting-point. From m to y lay off a number of equal distances, as eighths of an inch. Number them, and the scale is complete. If it be desired that 100 of these divisions should weigh 1 ounce, an ounce weight must be put in the patty-pan w, and the weight shifted along, until it balances; the intervening space must, then, be divided into 100 equal parts. It is desirable to construct two; one with a large weight, that will weigh 8 or 10 ounces; one with a small weight, to weigh the $\frac{1}{100}$th part of an ounce. This may be of wood, with a wire staple, about the thickness of a patent short-white pin. Nothing can be more convenient than these steelyard scales, as one weight answers for all, and never drops, and gets lost. It is well to have even a third steelyard scale. This may be made very thin, with a piece of wire, bent to the shape of fig. 171, for a weight. It may be made to weigh only the fifth part of an ounce; so that if divided into 100 equal parts, it will weigh the $\frac{1}{500}$th part of an ounce; so that in trying a new colour, as little can be mixed as will charge only one or two pill-boxes. A little notch should be made, at every division, with a fine, triangular file, to prevent the weight from slipping: every tenth division can be numbered; this is best done with a blacklead pencil, as ink runs; and on the end s, fig. 137, mark how many to the ounce. Possibly the trade, if desired, would make them to order. There is an instrument, somewhat similar, the chondrometer, for estimating the quality of grain; only this is constructed upon the principle of the log; the leverage and counterpoise showing the weight of a pint, but indicating that of a bushel; as the half-minute sand-glass times the running out of the knots of the line, and indicates nautical miles. Goods-weighing-machines, at the railways, are on the steelyard, or shifting leverage principle; also the machines warranted to furnish you with "your correct weight."

MONTGOLFIER BALLOONS.

To project a pattern gore. With a radius of 5 inches, fig. 139, from the point a, on the line b o, describe a semicircle; divide the upper quadrant into three equal parts, in the points g, e, c. Carry the same distance, once down, to k. Draw the lines f g, d e, a c, h k, parallel to each other, and at right angles to the line b o. Draw the line a k, and prolong it to s, so that the distance k s shall be 10 inches, *viz.*, twice the radius. From s, with the radius s k of 10 inches, describe the arc k n p. Draw o s parallel to h k; and, halfway between, m n.

As the distance from a to c is 5 inches, twice this will be 10, for the equatorial breadth of the gore; and, as it is intended to have 12 sheets of tissue paper (3 of each colour, yellow, blue, green, and red), and, consequently, 12 gores, 12 times 10 = 120, the half of which is 60, for the semi-circumference. Tissue paper is a very thin double-crown, 30 by 20. Lay down a straight line, b x, fig. 140, 60 inches in length, and divide it into 6 parts of 10 inches each; and draw parallels f g, d e, &c., at right angles to the meridian line b x. Measure the distance f g, fig. 139, and make f g, fig. 140, twice as much, half on each side; the same with d e, &c. Draw lines, connecting the extremities, till the fig. 140 is complete.

Instead of taking the trouble to construct fig. 139 with the compasses, the following measures will describe the pattern gore, fig. 140, at once; f g will be 5 inches, half to the left, half to the right, of the line b x; d e = $8\frac{2}{3}$ inches; a c = 10 inches; h k = $8\frac{2}{3}$ inches; m n = 7 inches; and o p = $6\frac{3}{4}$ inches.

These numbers are chosen to suit the size of the paper, to get the gores as large as possible, with the least waste: 60 inches are selected as the length of the gore, being twice the length of a sheet; 10 inches are $\frac{1}{6}$ the length, and exactly half the breadth of the sheet; so the length and breadth are both employed, without the slightest waste.

The pattern gore should be of cartridge or imperial brown.

To Cut the Gores.

Lay the twelve sheets of tissue paper flat upon each other, as at fig. 141; cut the pattern gore into two pieces, along the line a c, fig. 140, and lay them on the tissue paper, fig. 141. Mark round them with a blacklead pencil, and cut them through with a strong pair of scissors.

To Join the Halves Together.

Lay the bottom part of the gore, fig. 142, flat upon the table, and the top part upon it, so as to leave ¼ of an inch along the edge exposed, as shown by the shaded part. Paste this; and, without removing either piece, bend the pasted part of the lower half gore upon the top, and press it smooth down.

To Paste Two Gores Together.

Lay one gore with the point towards the *left* hand, and another upon it, a quarter of an inch back, as in fig. 143. Paste the shaded part of the lower gore, bend it over, and press as before. One of the gores, having been drawn in by pasting, is now narrower than the other. Finish the six pairs, and lay them by to dry.

To Paste the Pairs Together.

Lay one pair with the points towards the *right* hand, thus keeping the widest gore upwards; and another upon it, in the same manner, as fig. 144. Bend the top gore back upon itself, by folding it down the middle, as shown by the bottom shaded part, and lay a book or weight upon it, to keep it out of the way; paste, and double, as before; and so proceed, till the whole twelve are finished, and lying upon each other. Be careful to disturb none of the gores.

To Make the Last Joining.

Turn the points towards the *left* hand: lift up the top, and bend the ten inner gores back upon themselves: draw the top gore over, so as to make it lie upon the bottom. Paste as before. This is a somewhat troublesome operation, and it is advisable to have the assistance of a second person.

To Wire the Balloon.

Select a coil of iron wire, a little thicker than a pin; and remove the elasticity, as recommended with coloured gerbes. Before applying it to the balloon, practise the method of making a joining, as shown by fig. 145, simply bringing the ends together, crossing them about an inch from the extremity, and winding them tight round each other. Now lay the balloon, as in fig. 146, with a book, or a flat weight, upon it, to keep it down: curve the wire to the shape of the bottom of the balloon, and lay it about half an inch from the edge. Notch the paper, at the joinings of the gores; paste, and fold back. When perfectly dry, not before, turn the balloon over: cut the wire 2 inches longer than to meet; bend the ends together, and make the joining; paste, and bend over, the remaining half; and shape the wire to a circle. A star, of double-crown, may be pasted on the top, and a thread passed through it, to hold it by. To effect this, open the balloon, pass a book up it, spread the top flat on the book, and paste on it a star, or a circular piece of paper, the size of a penny. If a hole gets torn in it, it can be repaired in the same way.

Take two straight pieces of wire, a little longer than the diameter of the mouth of the balloon; pass them, half way, through a piece of sponge, at right angles to each other; fasten these wires across the mouth; saturate the sponge with wood naphtha, or methylated spirit, taking especial care that not a drop falls on the balloon; get someone to set fire to some pieces of paper on the ground; hold the balloon at a good height above the flame, to prevent the spirit, in the sponge, from catching. Three persons ought to assist in the start: one to manage the burning paper; two to hold the balloon steady, and to keep the mouth open, for the hot air to inflate it. As soon as the balloon gets expanded, set fire to the spirit; and directly it begins to pull against the hand, which it will do, let it go. A balloon of this size will require an ounce of spirit, and as much sponge as will absorb it. See fig. 167.

To project a gore for 24 sheets of tissue paper, let the radius be $7\frac{1}{2}$ inches; for 48 sheets, 10 inches. The shape of the balloon, when expanded, will be the same as fig. 139, rotating on the axis b o. If the pattern gore be projected from fig. 147, the balloon will be pear-shaped, and may be filled with gas, like those started from the Crystal Palace. The paper, 16-lb. double-crown, must have a coat of boiled linseed oil. The bottom should be open, and fitted with a cylindrical neck, about an inch diameter, and 3 inches long, made of writing paper of three thicknesses. Fig. 168 shows the magnesium light attached.

A flat weight may be constructed by making a thin deal box, the size and shape of a lucifer-match box; fill it with melted lead; nail a lid on it, and cover it with pasted paper. The same will serve for a paper-weight.

A spindle, on which to fire wheels, is shown at fig. 148; it is furnished with a fly-nut, which can be taken off, and put on, without trouble. As there is considerable difficulty in obtaining an article of this kind, its place may be supplied by a carriage-bolt, of which hundreds, if necessary, may be procured at the ironmonger's, of lengths varying from 1 inch to 12, or more. One of 6 inches will be suitable for most purposes. As the nut is small, and would be difficult to find, if dropped on the ground, on a dark night, proceed as follows. Turn a piece of beech, about $1\frac{1}{2}$ inch square, and $2\frac{1}{2}$ inches long, into a cylinder, with hemispherical ends, and a hole through it, like a popgun. Cut it, while yet in the lathe, or saw it across into two pieces, one an inch long, the other, an inch and a half. In the flat end of the inch length, mortise a hole, and sink the nut in it; take a circular piece of wood, with a hole bored in the middle, and screw it flat on, over the nut. Fig. 150, a, shows the turned piece; b, the circular piece, screwed on, to keep the nut from falling out. This serves the purpose of the fly-nut of fig. 148. The carriage-bolt is shown, first at fig. 149; then at fig. 150, with the nut screwed on. Take a piece of deal, fig. 155, about $1\frac{1}{2}$ inch square, and 20 inches long: make a hole through at a, and drive the carriage-bolt from behind; the shoulder, being square, will hold firm; in front, at c, slip on the above turned piece of wood, an inch and a half long: the piece, with the nut embedded in it, is, of course, put on at the end, b. The hole through this inch piece should be $\frac{1}{8}$ of an inch, or more, larger in diameter than the female (or counter) screw of the nut, in order that it may slip on and off with ease. At the top fix a strong wire, d, and file the end round and smooth: at s and x, in the side, fix two screw-eyes, or staples, to receive rocket sticks. At m and n bore two holes, through which it can be screwed to a post. The screw-eye at s ought to be about level with the top of the post. If there is too great a space between w and z, pieces of a roman candle case may be slipped on; but a popgun, sawed into short lengths, is better. Several should be kept in readiness, of different lengths, to suit the naves of different wheels. A piece of iron rod, about half an inch long, and to fit the hole, ought to be driven up furilonas, &c., to rest on the top of d, fig. 155; and the wire should be a little longer than the hole, that the piece may play clear of the upright, the top of which should be hemispherical, as drawn. Fig. 151 is a screw-ring; 152, a screw-eye: 153, a screw-hook; 154, a cup-hook. If a little bit, s, of 152 be sawed, or filed out, it makes a stronger cup-hook than the brass, for match-weights.

WINGED ROCKETS.

An ⅜ rocket is the smallest size worth making: it is, also, sufficiently large; and, as it requires a machine to fire it from, it is best to keep to one size. The head, fig. 156, may be the same as the external diameter of the case, or it may be slightly enlarged: to receive a parachute, it may be 12, or even 16 inches high; and the top may be closed with a lid, fig. 163, formed of a case choked tight; or by simply stopping it with a piece of turned wood, or with a thin bung, pushed in, as a cork into a bottle. A bung may be readily sawed thin with a fine-toothed saw; if required to be made smaller in diameter, it must be cut round sidewise, not longitudinally. If the head is to receive stars, it must not be above 8 inches high. To construct the wings, take a thin piece of deal, or tin, and cut out a right-angled triangle, fig. 157. Make a b = 7½ inches, b c = 1½ inch. Take a piece of imperial brown paper, 8 inches long, 4 broad, and fold it down the middle lengthwise, so as to become 8 by 2. Lay the sloping edge, a c, of fig. 157 along the folded edge of the paper, a c, fig. 158, and mark round it, with a pencil. About ⅜ of an inch distant, draw the line d e parallel to a b. Cut through with the scissors, and the paper will be of the shape of fig. 158. Bend the shaded parts flat, to form a flap, or kind of hinge. Paste the surface of the two triangles, and press them together to form a double thickness. If this does not make the wing stiff enough, insert, between them, a triangular piece of cardboard, to make it of three folds, like a shirt-collar. Take a piece of sheet tin, or a piece of cardboard, fig. 161. Let m n be 7 inches; and m p such that it shall exactly wind round the rocket. Divide it into 3 equal parts, by the parallel lines r and s, and at each end cut out little pieces, as shown at a a, &c. Now, if this be wound round the rocket, marks can be made through the slits a a, with a blacklead pencil, or with a stencil-brush, dipped into blacklead powder, such as used for polishing grates. These marks will indicate the places for the wings. Paste or glue them on, and secure the joinings with another slip of paper pasted over them, as shown by the dotted lines, beside the wings, fig. 156. The wings will branch out at angles of 120° divergence from each other, like the lines a a, b b, c c, fig. 162. The rocket is thus winged; and, so far, complete.

To Construct a Slot-Tube, or Rocket-Guide.

Procure 3 pieces of planed deal, ½ an inch thick, 2 inches broad: let one piece be 6 feet 3 inches long; the other two, each 6 feet. Get a smith to make 4 triangular iron holdfasts, k m n, fig. 162, 5 inches along each side; the iron may be a trifle above ⅛ of an inch thick, and ⅝ broad: in each side, let two holes be made, as in the usual holdfasts, to receive screws. The screws are represented at z z, fig. 162, which is a cross section of the tube. Fasten the 4 iron holdfasts round the 3 pieces of deal, as represented by w w, fig. 160, at equal distances. The pieces of wood will thus form a triangular spout, with open corners. The pieces are to be level at the top; the longest one, of 6ft. 3in., will, thus, be 3 inches below the other two, as at r r, fig. 160. At s s bore two holes, about ⅛ of an inch diameter. Take a small piece of deal, fig. 159, and fix in it two pieces of wire, so as to have the appearance of a tuning-fork. These wires are to go through the holes s s, of fig. 160, to form a resting-place, or support for the rocket, after it has been pushed up the tube. At the top of one of the short pieces, is to be a hook, k, by which to suspend it from a post. This post ought to be 9 or 10 feet high, that the rocket may be fired without having to stoop. Everything being ready, the rocket is to be put up through the bottom of the tube, and the wires of fig. 159 pushed through, for it to rest upon. It may then be fired.

These rockets are peculiarly fitted for asteroids, as, from their lightness, they rise to great heights, with immense rapidity, not in a continuous curve, of uniform flexure, but with a sweeping serpentine motion, as indicated by the line, fig. 164. They might be employed, with great advantage, for distress rockets at sea.

They could also be made with four wings, and fired in volleys, from batteries constructed with square tubes. See the design, fig. 165.

LIST OF PRICES.

- Ammonia, Muriate, or Salammoniac, 6*d.* per lb.
- Antimony, Sulphuret, 6*d.*
- Arsenic, Red Sulphuret, or Realgar, 1*s.*
- Yellow Sulphuret, or Orpiment, 1*s.*
- Barytes, Chlorate, 6*s.*
- Nitrate, 8*d.*
- Cast Iron Borings, 4*d.*
- Charcoal, 3*d.*
- Coal tar naphtha, or mineral naphtha, 4*d.* per pint.
- Copper, Black Oxide, 3*s.*
- Oxychloride, 4*s.*
- Sulphuret, 4*s.*
- Dextrine, 1*s.*
- Gunpowder, F, 8*d.*;
- FF, 9*d.*;
- FFF, 10*d.*
- Meal, 9*d.*
- Lamp Cotton, 2*d.* per oz.
- Lead, Chloride, 2*s.* 6*d.*
- Nitrate, 1*s.*
- Protoxide, or Litharge, 6*d.*
- Deutoxide, Minium or Red Lead, 6*d.*
- Litmus, 2*d.* per oz.
- Magnesium Filings, 7*s.* per oz.
- Mercury, Chloride, or Calomel, 6*s.*
- Sulphuret, or Æthiop's Mineral, 6*s.*
- Methylated Spirit, 6*d.* per pint.
- Oxalic Acid, 1*s.*
- Plaster of Paris, 3*d.* per packet.
- Potash, Chlorate, 1*s.*
- Nitrate, Nitre, Saltpetre, Sal-prunella, 6*d.*
- Shellac, in Flake, 1*s.* 3*d.*
- in Powder, 2*s.*
- Soda, Bicarbonate, 4*d.*
- Oxalate, 2*s.*
- Stearine, Composite Candle, 1*d.*
- Steel Filings, 8*d.*
- Strontian, Carbonate, 3*s.*
- Nitrate, 8*d.*

- Sugar, 5*d.*
- Sulphur, Flowers. Sublimed Sulphur, 4*d.*
- Vegetable Black, 10*d.*
- Wax, 1*d.* per cake.
- Wood Naphtha, 7*d.* per pint.
- Zinc Filings, 3*s.* per lb.

CONCLUDING REMARKS.

Sublimation is the volatilization of solid substances by heat, and their crystallization by cold again into solids.

The products of sublimation (sublimates) have received the name of flowers from their soft efflorescence, or aggregation of minute spicular crystals into flakes; as flowers of sulphur, the crystallized refrigerated vapour of burning brimstone; flowers of benzoin, benzoic acid; corrosive sublimate, bichloride of mercury; sublimed arsenic, camphor, sal-ammoniac; vegetable and lamp black, the condensed fumes of burning oils and resins; soot, the flakes deposited in chimneys from the smoke of burnt wood and coals.

Distillation is the evaporation of liquid substances by heat, and their condensation by cold again into liquids.

The products of distillation (distillates) are usually termed spirits; as spirit of wine, alcohol or brandy; spirit of grain, gin, hollands, or whiskey; spirit of molasses, rum; spirit of naphtha; benzine, &c.

Water heated and cooled, combines in resemblance the effects of sublimation and distillation; aqueous vapour by congelation crystallizing into snow; and by condensation liquefying into water.

The condensation of steam into water is familiar to everyone. It is stated that in St. Petersburg, upon the sudden admission of a current of cold air into a crowded assembly-room, the vapour in the air was immediately congealed, and fell in the form of snow flakes. Probably snow might be produced artificially by driving steam into a vessel preparatively cooled below the freezing point.

Gums are the exudation of trees, vegetable mucilage thickened by exposure to the atmosphere; as gum from cherry and plum trees; gum arabic, from varieties of the acacia, Turkey, East India, Senegal, or Barbary; Turkey gum arabic is the best.

Resins are the exudation of trees, generally evergreens, essential oils inspissated by oxygenation: as mastic, sandarac, benzoin.

Gums are soluble in water; resins in alcohol and essential oils.

Gums dry and swell up by heat; resins soften and melt.

Gum resins are partly resinous and partly mucilaginous; as lac, assafœtida, galbanum. In submitting shellac to the action of alcohol, the whole is never entirely dissolved; as the lac contains, besides the resin, a mucilage which floats about in the liquid, and renders it turbid.

Many substances which go under the name of gums in commerce, are in reality resins or gum-resins.

Native turpentine, the juice of trees of the fir tribe, of the consistency of honey, yields on distillation spirit of turpentine, called also oil of turpentine, and by painters turps; the dry mass left behind in the retort is colophony or rosin. Rosin is soluble in alcohol, and is therefore a resin; rosin and resin, however, are not synonymous; all rosin is a resin; but all resins are not rosin. Rosin has been tried in pyrotechny, but is of no use: a solution of it in spirit will bind stars; but it renders them white and smoky.

Volatile, ethereal, or essential oils are obtained from plants by distillation with water; as oil of roses, lavender, thyme, peppermint, aniseed, &c.

Fixed oils are obtained from animal fat by heat; and from seeds of plants by pressure and percussion; as, train oil, cod-liver oil; palm oil, croton, linseed, cottonseed, &c.

Oxychloride of copper, if difficult to procure, may be made by laying thin pieces of copper in a dish, and pouring upon them a mixture of half water, and half hydrochloric or muriatic acid. The next day remove them, and lay them on a board in the shade to dry. When dry, brush off the green powder which will be found on the outside, with a toothbrush, into a basin of water. After a quantity is obtained, wash it as directed for sulphur, and dry it in the bag, fig. 33. Test it with litmus paper to ascertain if free from acid.

Saw a piece of coke or charcoal in two, and on the flat surface place a few copper filings; direct upon them the flame of a lamp or candle with the blow-pipe; they will simply become red-hot. Lay a few more filings, and on them a little calomel or sal-ammoniac: now direct the flame, and a beautiful blue colour will be produced. Any of the salts of copper may be used to obtain the same effect, the chlorine gas liberated from the calomel (chloride of mercury) or from the sal-ammoniac (ammonic chloride) giving a blue colour to all preparations of copper burnt in it.

Weights for quickmatch may be made by nearly filling the brass tube with useless rusty old nails, tacks, screws, or odd bits of iron, or brass; then pouring in melted lead. If the ladle will not hold enough lead to fill it at once, it may be poured in at twice, thrice, four or more times. A tube 1¼ inch diameter and 6 inches long will weigh 2½ lb. This will keep a great length of match tight and straight. Half-an-inch at each end of the tube should be solid lead, one to receive the screw, and one to make a firm bottom.

If at any time the basil end of a pinwheel pipe should be too small to admit the nose of the funnel, it may be enlarged by binding a gum strip round it. If pinwheels are too dry, they break in winding, from the hardness of the composition; if too damp, from the softness of the paper. As paper cannot be relied upon for being always of uniform thickness, if it be found that a pinwheel pipe is too thin, cut the strip a little broader. Discretion may be used in all cases.

Coloured lances may be primed with meal powder very slightly damped with thin lac solution. Leader pipes may, if preferred, be fastened to lances with patent short whites: they may be procured at the haberdasher's; the price of them is 2*d.* per oz. Push the pin through the side of the leader, down into the side of the lance; then make a triangular hole through the middle of the leader, down into the middle of the lance, turning the tool round to break the priming, and secure the leader with a gum strip. The gum strip, bent round, assumes the shape of the capital letter **T**. The best tool for making the holes is a steel bradawl, ground triangular and to a sharp point; another bradawl ground to a tapering point like a needle, about ¾ of an inch long, may be used for making holes up the lances to receive the wires. Scissor-grinders will shape them, if you have not a grindstone; or they may be rubbed on a stone, such as used by mowers to whet their scythes. Afterwards set them sharp on a hone.

Fig 169 is a wasp-light. The proper composition will be found among the fuses: drive it into a roman candle case with the rammer, fig. 4. Bring the leader from the mouth b, backwards along the outside of the case, and tie it in a couple of places, as drawn. Evening is the best time to use it: push the end b into the nest, light at a, and retire.

Instead of a five-pointed star, a seven-lance star, fig. 170, may be employed. To form it, have a piece of deal board, half-an-inch thick, 6½ inches square: draw the diagonals, to find the centre; and, with a pair of compasses, stretched to a 3-inch radius, describe a circle. Carry the radius 6 times round it; and in the points and the centre drive 7 French nails; cut off their heads, and fix on them 7 lances: the middle one, crimson; the others, 3 green and 3 blue, placed alternately.

In forming a rocket spindle, taper it no more than will just make it deliver: the thicker it is left at the top, the stronger of course it will be. For small rockets, $\frac{3}{8}$ or $\frac{1}{8}$, a brass, iron, or steel wire, with a few notches filed in it, or made jagged with a cold chisel and hammer, driven into a block, will hold firm without a screw. I have seen them driven into a piece of thick plank, and the nipple formed with an inch-length of wood, cut cylindrical, bored and slipped over the spindle, like c in fig. 155. Indeed, a spindle might be formed from a 5-inch or 6-inch carriage-bolt. At Woolwich Arsenal rockets are charged solid. The fuse is shaped into pellets, something like large peppermint lozenges, or cylindrical cakes of paste blacking, by hydraulic pressure in a mould. These pellets, discs, or cakes, which are almost as hard as a stone, are put into the case, and pressed in tight: the rocket is then fixed upright, and slowly drilled, as I have seen, with a conical borer, working vertically, to let the dust fall and clear itself. This mode must not be imitated by an amateur; indeed, without accurate machinery, the desired object could not be effected, and there is constant liability to danger.

The fuse of a rocket, when consolidated, assumes the form of fig. 18, with the head sawn off, except that the hollow is tapering, instead of cylindrical; and the rocket stands thus—

$\frac{1}{3}$ cup + 6 choke & hollow + $1\frac{1}{3}$ solid + $\frac{1}{3}$ plaster = 8 diameters.

In the trade, meal powder, saltpetre, and charcoal, go by the names of meal, petre, and coal. Common coal, for burning in fires, is never employed in pyrotechny; it would produce only dull red sparks and smoke.

Meal, or petre, added to a fuse quickens it; sulphur slackens it. 6 meal, 1 sulphur, make a quickmatch that blows through a leader with great violence. 1 meal, 1 sulphur, will scarcely burn; pure meal only should be used for match, or grain powder with hot starch. It has already been stated that nitre in powder is sometimes adulterated with salt, and that it is impossible to make a rocket with such stuff. Powdered chlorate of potash is sometimes adulterated with nitre: with such mixture it is equally impossible to produce good colours: nitre whitens flame, and overpowers colour.

Chlorate of potash, charcoal, sulphur, stearine, used separately, with discretion, vivify colours; calomel deepens the colour, but slackens the flame.

Star compositions which inflame vigorously in dry summer weather, will often scarcely burn at all in damp weather; this is especially the case with stars containing nitrate of strontian.

In washing sulphur, stir it with a wooden spoon; if a silver one were used, a black sulphide would be formed on the surface, very difficult to remove. If silver coins in the pocket get tarnished while using sulphur, rub them with salt, or chalk, or whiting.

A magnet is convenient for lifting tacks, small screws, &c., from divisions in nail boxes.

If the brass tube formers get tarnished, scrape off the lacquer with a knife, sand paper them anew, and give them a fresh coat of lacquer.

Let all wheel frames and woodwork be coloured black, either with paint, or with a mixture of vegetable or lamp black and size, or thin glue, to prevent their being seen. A white thread hung upon a bush, is visible many yards off; a black one can scarcely be seen a few feet distant. Black is not only invisible, but it throws the brilliancy of sparks, and the vividness of colours, into stronger relief.

Let every article be dried, reduced to a fine powder, put into a clean bottle, and carefully corked: also let every bottle be labelled: the labels are best stuck on with paste, not gum: gum labels are apt to drop off in damp weather.

Let all metallic articles, liable to rust, be wiped with a rag dipped into olive oil, before being laid by for future use.

Before putting aside the six-inch circular frying-pan, set it over the fire till warm, put into it a lump of tallow, and smear it with a rag: when wanted for use, set it on the fire, put into it a cupful of water and a piece of soda; make the water boil, and stir it well round; pour away the water, and dry the pan over the fire.

Let muslin sieves always be dried before being put aside; also, again, before use. Zinc sieves may simply be wiped dry.

Have a place for everything; and keep everything in its place.

Faraday, the great master in experimental lectures, always devoted many hours to the preparation of his experiments for each lecture. No point, however trifling, bearing upon the success of the experiment, was considered unimportant: he used to try the stoppers of all the bottles he had to use, to see that they had not become fixed, and thus would cause delay by requiring forcible opening. His example cannot be too carefully copied. Before firing a display, all posts, spindles, lines, staples, screws, touch-paper, portfires, pieces of leadered quickmatch, &c., should be carefully provided. A yard of tape slowmatch, hung to a nail at the top of a post, will supply fire for a long time.

A book should be kept for future guidance, in which should be written the quantity of composition required to make a certain number of articles of a certain size: by attending to this, much waste will be prevented.

Never, upon any account, leave compositions lying about; and let nothing be done by candlelight, except making cases. Quickmatch, especially, ought to be kept locked up, so that nobody can get to it.

Never put squibs, crackers, &c., into the pockets: a stray spark might ignite the whole, and cause most serious mischief.

FUSES.

ROMAN CANDLE.

Number	1	2	3	4	5	6	7	8	9	10	11	12	13	14
Sulphur	4	2	3	1	1	2	7	3	4	6	8	3	8	-
Charcoal	3	3	3	2	1	3	8	1	1	7	9	3	11	2
Nitre	8	2	8	4	3	9	21	4	5	18	18	10	32	1
Meal powder	8	8	3	3	2	4	12	5	4	4	4	7	-	3

ROCKET.

Number	1	2	3	4	5	6	7	8	9	10	11	12	13	14
Sulphur	1	1	12	4	8	4	2	4	2	2	1	1	8	1
Charcoal	4	2	17	5	11	7	4	8	12	8	2	2	27	2
Nitre	8	5	50	16	32	16	9	16	20	16	4	4	36	4
Meal powder	-	-	-	-	-	-	-	3	1	1	1	2	6	2
Steel-filings	-	-	-	-	-	-	-	-	-	-	-	-	-	1

WHEEL AND FIXED CASE.

Number	1	2	3	4	5	6	7	8	9	10	11	12	13	14	15	16
Meal powder	8	24	8	36	4	18	8	12	42	4	16	10	13	16	20	40
Sulphur	1	1	-	1	-	-	1	1	3	-	-	-	1	-	1	4
Charcoal	1	4	1	4	1	-	-	-	-	-	-	-	-	-	-	3
Nitre	2	3	-	-	2	2	1	3	8	-	2	-	1	-	-	24
Steel-filings	-	-	-	-	-	5	3	3	5	1	5	-	-	-	-	6
Vegetable black	-	-	-	-	-	-	-	-	-	-	-	1	2	-	-	1
Realgar	-	-	-	-	-	-	-	-	-	-	-	-	1	-	-	1
Litharge	-	-	-	-	-	-	-	-	-	-	-	-	-	3	3	2

SQUIB AND SERPENT

OILED TAILED STARS FOR ROCKETS AND SHELLS.

Number	1	2	3
Charcoal	9	6	2
Sulphur	9	5	2
Nitre	32	18	9
Meal powder	24	12	5
Sulphuret of Antimony	16	9	4

To 1 ounce add twenty-four drops of boiled linseed oil: rub them thoroughly together in a mortar; then spread out the mixture, for a few days, to dry. When dry, mix with starch, dextrine solution, or gum water, and chop into ⅜ or ½ inch cubical blocks.

STEEL STARS FOR ROCKETS AND SHELLS.

Number	1	2	3	4
Nitrate of Lead	8	24	28	-
Chlorate of Potash	3	5	6	5
Charcoal	2	6	6	1
Steel-filings	2	6	6	3
Nitre	-	4	3	-
Shellac, fine	-	-	1	-
Sulphur, washed	-	-	-	1

Mix together on paper, damp with lac solution, and chop into cubical blocks. The composition may also be pumped into roman candle streamer stars.

No substance combines better with salts of copper, than sugar. Sugar, put into the bowl of a tobacco pipe, and placed in the fire, burns fiercely, and is converted into caramel. This, poured on to a plate, slightly smeared with butter, to prevent its sticking, hardens on cooling; and is used for colouring brandy, vinegar, gravy, porter, coffee, &c. Stearine must be scraped very fine from a composite candle.

BLUE STARS AND LANCES WITHOUT SUGAR.

Number	1	2	3	4	5	6	7	8	9	10	11	12	13	14	15	16	17	18	19	20	21	22
Chlorate of Potash	5	40	18	40	6	8	48	40	24	16	30	24	8	22	40	6	16	24	2	5	40	40
Calomel	4	20	8	28	2	2	12	12	6	8	10	8	1	8	20	3	1	8	-	-	-	-
Sulphuret of Copper	4	20	10	28	-	2	-	-	-	-	-	-	3	6	25	5	-	-	-	1	20	22
Shellac	1	5	-	-	-	-	2	2	1	-	-	1	-	-	5	-	-	1	-	-	5	-
Oxychloride of Copper	-	2	-	-	-	-	8	9	6	5	10	4	1	2	-	-	2	-	1	1	-	-
Dextrine	-	-	5	10	-	-	-	-	-	-	-	-	-	-	2	2	-	-	-	-	-	-
Sulphur	-	-	-	-	3	4	4	-	1	2	3	2	3	5	-	-	4	2	1	2	5	5
Stearine	-	-	-	3	-	-	1	1	2	2	3	1	-	-	-	-	-	-	-	-	-	3
Black Oxide of Copper	-	-	-	-	1	1	1	-	-	-	-	-	-	-	-	-	-	-	-	-	-	-
Copper-filings	-	-	-	-	-	-	-	-	-	-	-	-	-	-	-	-	-	5	-	-	-	-
Sal Ammoniac	-	-	-	-	-	-	-	-	-	-	-	-	-	-	-	-	-	-	-	-	6	6

CRIMSON AND SCARLET STARS AND LANCES.

Number	1	2	3	4	5	6	7	8	9	10	11	12	13	14	15	16	17	18	19	20	21	22
Chlorate of Potash	16	8	16	16	24	16	5	16	8	16	25	4	32	6	16	28	32	26	96	8	24	8
Nitrate of Strontian	16	16	32	32	-	16	4	24	5	6	30	7	48	5	-	-	42	10	72	12	18	12
Sulphur, washed	5	6	9	12	6	5	2	-	1	-	10	1	6	-	-	-	13	5	24	2	-	2
Charcoal, fine	1	1	1	1	1	-	-	-	-	-	-	-	-	-	-	-	-	-	-	-	-	-
Shellac	-	1	4	2	2	-	-	7	1	4	3	1	12	1	-	-	4	3	21	4	5	3
Calomel	-	-	7	-	6	1	-	14	2	2	9	2	12	5	-	12	12	10	42	-	-	4
Sulphure of Copper	-	-	3	1	-	-	-	1	1	1	3	1	-	1	-	-	4	1	4	1	5	-
Realgar, or Orpiment	-	-	-	-	-	1	1	1	-	-	-	-	-	-	-	-	-	-	-	-	1	-
Vegetable black	-	-	-	-	-	-	-	-	-	-	1	-	1	-	-	-	1	-	-	-	-	1
Loaf Sugar	-	-	-	-	-	-	-	-	-	-	-	-	-	-	7	12	-	-	-	-	-	-
Carbonate of Strontian	-	-	-	-	10	-	-	-	-	-	-	-	-	-	11	5	-	-	-	-	-	-

It is impossible to powder shellac sufficiently fine, by hand; and, twenty years ago, it could not be procured. About that time the drug-grinders, finding a demand for it, submitted it to the action of the stamping mills, (mechanical pestle and mortar), and now it can be obtained at most shops. Chertier mixed it with salt; melted the two together; powdered the mixture; and washed out the salt. Such process is needless now. It is useless, unless as fine as wheaten flour. Page, of 47, Blackfriars Road; and Chubb, of 29, Old Street, St. Lukes, London, supply it.

GREEN OR EMERALD STARS AND LANCES.

Number	1	2	3	4	5	6	7	8	9	10	11	12	13	14	15	16	17	18	19	20
Chlorate of Potash	16	8	132	32	144	8	3	1	-	16	6	12	16	3	48	22	24	108	24	16
Nitrate of Barytes	16	8	108	54	160	21	2	-	-	8	7	5	8	3	42	22	32	108	32	48
Chlorate of Barytes	-	-	-	-	-	3	2	2	4	8	3	4	4	-	-	-	-	-	-	-
Sulphur, washed	5	4	6	6	4	7	1	1	1	5	-	-	1	-	-	-	10	18	8	12
Charcoal, fine	1	1	-	-	-	1	-	-	-	1	-	1	-	-	-	-	-	-	-	1
Sulphuret of Antimony	-	1	-	-	-	-	-	-	-	-	-	-	-	-	-	-	-	-	2	-
Calomel	-	-	48	27	100	-	-	-	-	-	5	2	2	2	-	10	-	48	-	8
Shellac	-	-	-	24	12	12	2	-	-	-	-	3	4	-	7	1	-	24	-	5
Vegetable black	-	-	-	1	1	1	-	-	-	-	-	-	-	-	-	-	-	1	-	-
Loaf Sugar	-	-	-	-	-	-	-	-	-	-	5	-	-	2	7	14	-	-	-	-
Salammoniac	-	-	-	-	-	-	-	-	-	-	-	-	-	-	5	-	-	-	-	-
Orpiment, or Realgar	-	-	-	-	-	-	-	-	-	-	-	-	-	-	-	-	3	-	3	-
Sulphuret of Copper	-	-	-	-	-	-	-	-	-	-	-	-	-	-	-	-	-	-	-	2

GREEN OR EMERALD STARS AND LANCES.

Number	1	2	3	4	5	6	7	8	9	10	11	12	13	14	15	16	17	18	19	20
Chlorate of Potash	16	8	132	32	144	8	3	1	-	16	6	12	16	3	48	22	24	108	24	16
Nitrate of Barytes	16	8	108	54	160	21	2	-	-	8	7	5	8	3	42	22	32	108	32	48
Chlorate of Barytes	-	-	-	-	-	3	2	2	4	8	3	4	4	-	-	-	-	-	-	-
Sulphur, washed	5	4	6	6	4	7	1	1	1	5	-	-	1	-	-	-	10	18	8	12
Charcoal, fine	1	1	-	-	-	1	-	-	-	1	-	1	-	-	-	-	-	-	-	1
Sulphuret of Antimony	-	1	-	-	-	-	-	-	-	-	-	-	-	-	-	-	-	-	2	-
Calomel	-	-	48	27	100	-	-	-	-	-	5	2	2	2	-	10	-	48	-	8
Shellac	-	-	-	24	12	12	2	-	-	-	-	3	4	-	7	1	-	24	-	5
Vegetable black	-	-	-	1	1	1	-	-	-	-	-	-	-	-	-	-	-	1	-	-
Loaf Sugar	-	-	-	-	-	-	-	-	-	-	5	-	-	2	7	14	-	-	-	-
Salammoniac	-	-	-	-	-	-	-	-	-	-	-	-	-	-	5	-	-	-	-	-
Orpiment, or Realgar	-	-	-	-	-	-	-	-	-	-	-	-	-	-	-	-	3	-	3	-
Sulphuret of Copper	-	-	-	-	-	-	-	-	-	-	-	-	-	-	-	-	-	-	-	2

If powdered nitrate of barytes, and shellac crushed by being hammered in a bag, are mixed together, and melted in a pipkin, over the fire, the mixture, when cold, may be reduced to a powder in an iron mortar, with patience. Take Number 6. Weigh out 21 parts nitrate of barytes, and 2 parts of coarsely powdered lac; melt them together; when cold, powder them; and add the other substances in proper proportion. Shellac may be melted with nitrate of strontian, in the same way.

DEEP AND PALE YELLOW STARS AND LANCES.

Number	1	2	3	4	5	6	7	8	9	10	11	12	13	14	15
Chlorate of Potash	8	4	12	16	12	16	4	4	16	3	8	16	16	4	8
Oxalate of Soda	3	2	8	4	4	4	3	1	4	4	4	5	-	1	-
Bicarbonate of Soda	-	-	-	-	-	-	-	-	-	-	-	-	3	-	3
Nitrate of Strontian	-	-	-	-	-	-	20	-	-	-	16	-	4	-	-
Carbonate of Strontian	-	-	-	3	2	3	-	-	-	-	-	-	-	-	-
Nitrate of Barytes	-	-	-	-	-	-	-	-	4	10	-	-	3	-	-
Sulphur, washed	-	-	-	4	4	4	-	1	2	-	6	5	-	-	-
Shellac	2	1	3	-	1	1	5	-	2	3	1	-	4	1	-
Stearine	-	-	-	-	-	1	-	-	1	-	-	-	-	-	-
Charcoal, fine	-	-	-	-	-	-	-	-	-	-	1	1	-	-	-
Orpiment, or Realgar	-	-	-	1	1	1	-	-	-	-	-	-	-	-	1
Sugar	-	-	-	-	-	-	-	-	-	-	-	-	-	1	3

MAUVE AND LILAC STARS AND LANCES

Number	1	2	3	4	5	6	7	8	9
Chlorate of Potash	28	17	60	40	25	24	24	25	12
Calomel	12	-	-	-	10	12	12	-	-
Shellac	4	-	-	-	5	5	5	-	-
Nitrate of Strontian	4	4	25	14	-	4	-	16	16
Sulphuret of Copper	2	7	20	-	5	2	2	-	-
Stearine	1	-	-	-	1	-	1	-	-
Sulphur, washed	-	7	35	16	-	-	-	12	2
Chloride of Lead	-	1	-	2	-	-	-	-	-
Nitrate of Lead	-	-	-	-	10	-	12	1	-
Oxychloride of Copper	-	-	8	12	-	-	-	6	-
Salammoniac	-	-	1	-	-	-	-	-	-
Vegetable black	-	-	-	1	-	-	-	1	1
Nitre	-	-	-	2	-	-	-	2	1
Carbonate of Strontian	-	-	-	-	5	-	4	-	-
Orpiment, or Realgar	-	-	-	-	-	-	1	-	1

PURPLE AND VIOLET STARS AND LANCES.

Number	1	2	3	4	5	6	7	8	9	10	11	12
Chlorate of Potash	42	28	48	16	6	16	3	6	26	30	96	24
Nitrate of Strontian	42	14	48	-	4	-	-	1	-	-	24	-
Sulphur, washed	13	-	28	2	1	6	1	3	-	3	-	2
Calomel	12	14	28	7	2	6	2	2	20	8	48	8
Sulphuret of Copper	4	1	40	8	1	-	3	-	3	12	1	-
Shellac	4	5	1	-	1	-	-	-	-	4	-	1
Vegetable black	1	-	-	-	-	-	-	-	-	-	-	-
Black Oxide of Copper	-	-	-	-	-	4	4	1	-	-	-	-
Carbonate of Strontian	-	-	-	-	-	-	-	-	4	12	-	-
Sugar	-	-	-	-	-	-	-	-	14	-	42	-
Oxychloride of Copper	-	-	-	-	-	-	-	-	-	-	-	4
Stearine	-	-	-	-	-	-	-	-	-	-	-	2

MAGNESIUM COLOURS FOR STARS AND ASTEROIDS.

Colours.	Crimson.	Scarlet.	Green.	Blue.	Yellow.	White.
Nitrate of Strontian	8	6	-	-	-	-
Chlorate of Barytes	-	-	12	-	-	-
Oxychloride of Copper	-	-	-	2	-	-
Oxalate of Soda	-	-	-	-	2	-
Sulphuret of Antimony	-	-	-	-	-	1
Chlorate of Potash	2	4	-	5	4	-
Nitre	-	-	-	-	-	12
Sulphur	2	2	1	2	-	4
Charcoal	1	-	-	-	-	-
Shellac	-	2	3	1	1	-
Calomel	-	4	-	2	-	-
Magnesium-filings	2	3	2	2	1	2

SLOW FIRES, TO BE HEAPED UPON A TILE IN SHAPE OF A CONE, AND LIT AT TOP.

Colours.	Red.			Green.			Purple.		Yellow	
Nitrate of Strontian	16	24	108	-	-	-	108	72	20	-
Nitrate of Barytes	-	-	-	16	16	16	-	-	-	10
Oxalate of Soda	-	-	-	-	-	-	-	-	3	5
Sulphuret of Copper	-	3	30	-	-	-	24	3	-	-
Chlorate of Barytes	-	-	-	-	-	12	-	-	-	-
Chlorate of Potash	1	3	12	1	1	-	9	4	2	2
Charcoal, fine	1	-	-	1	-	-	-	-	-	-
Calomel	-	6	24	-	5	9	24	18	-	-
Sulphur, washed	4	8	39	4	2	7	39	24	4	1
Shellac	-	1	2	-	2	1	2	3	2	6
Vegetable black	-	-	1	-	-	-	1	2	-	-

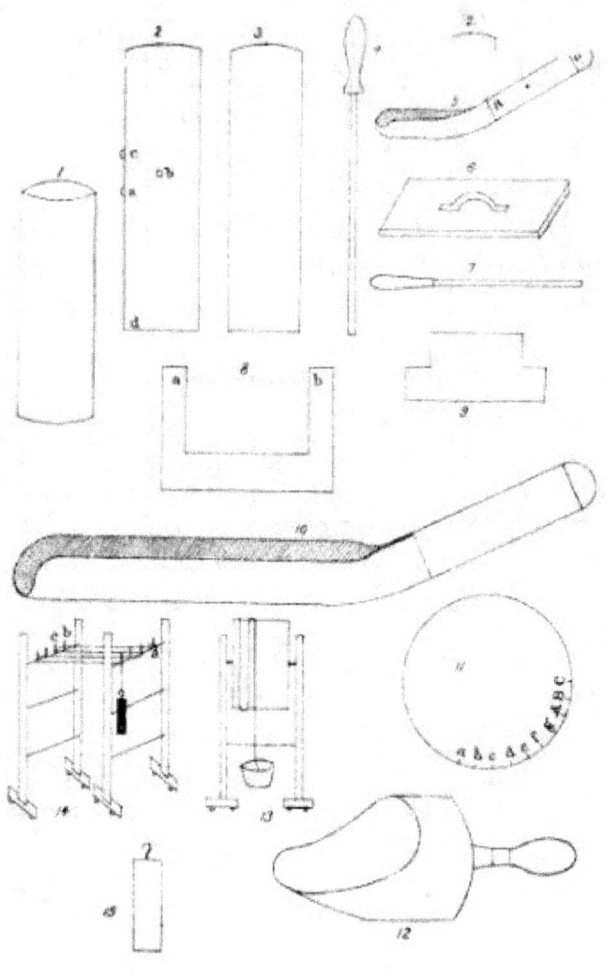

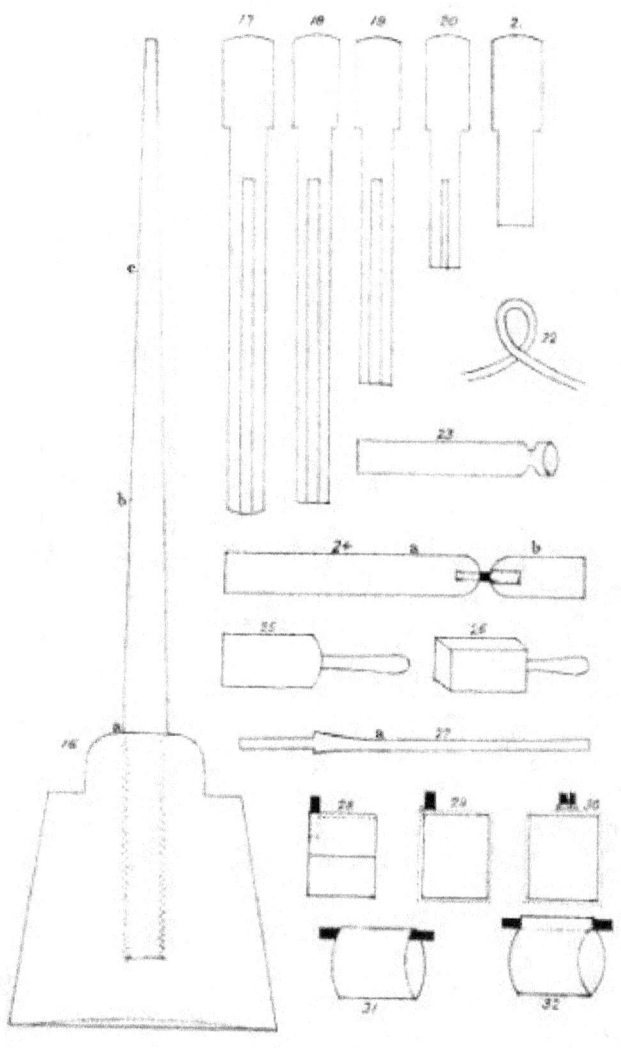

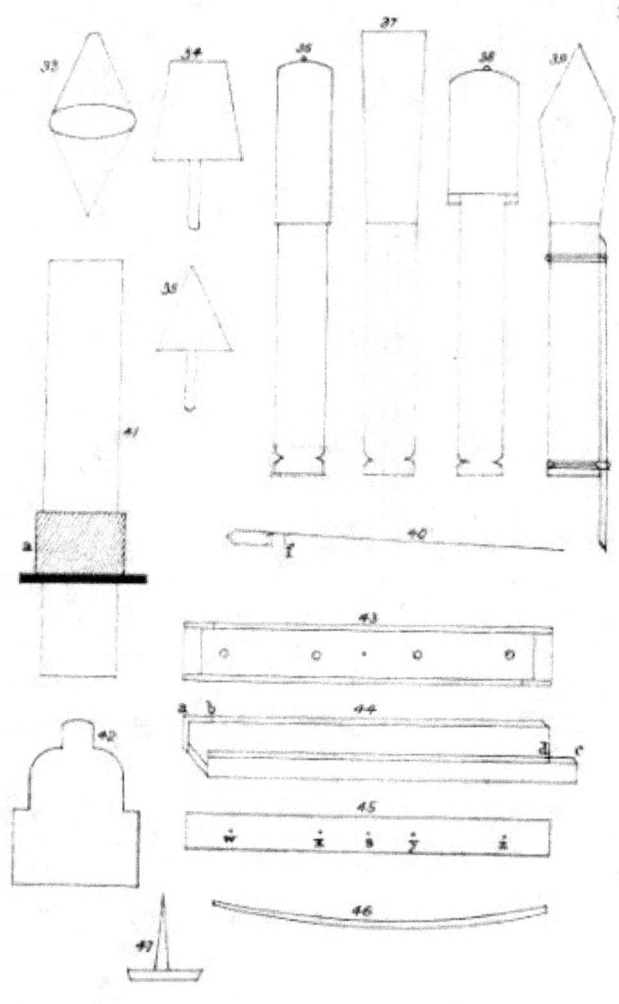

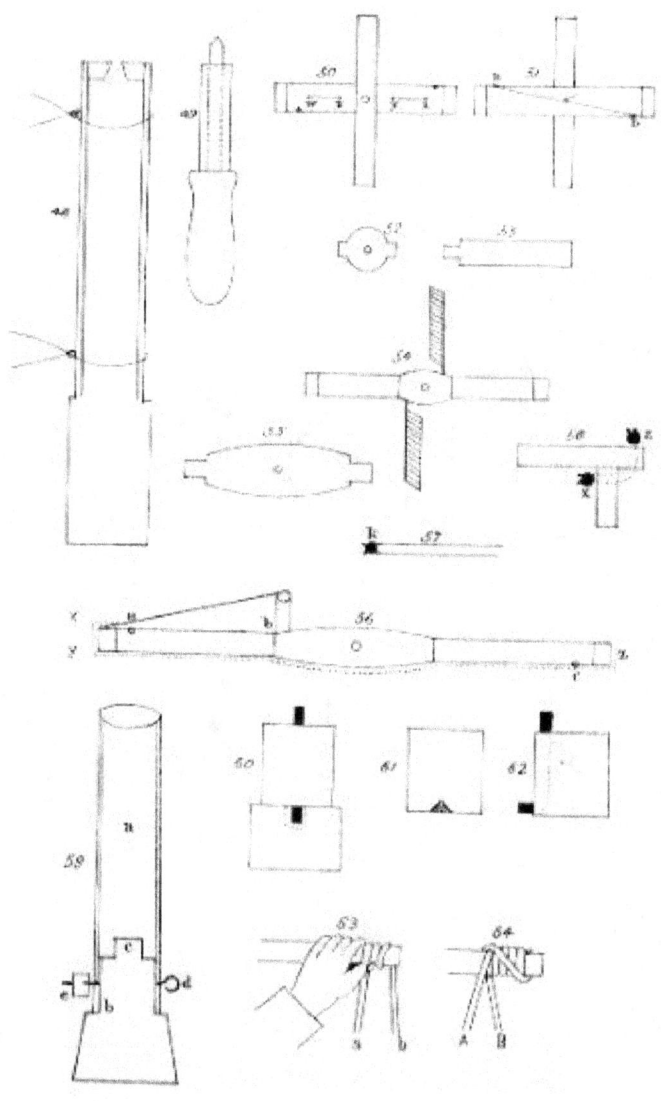

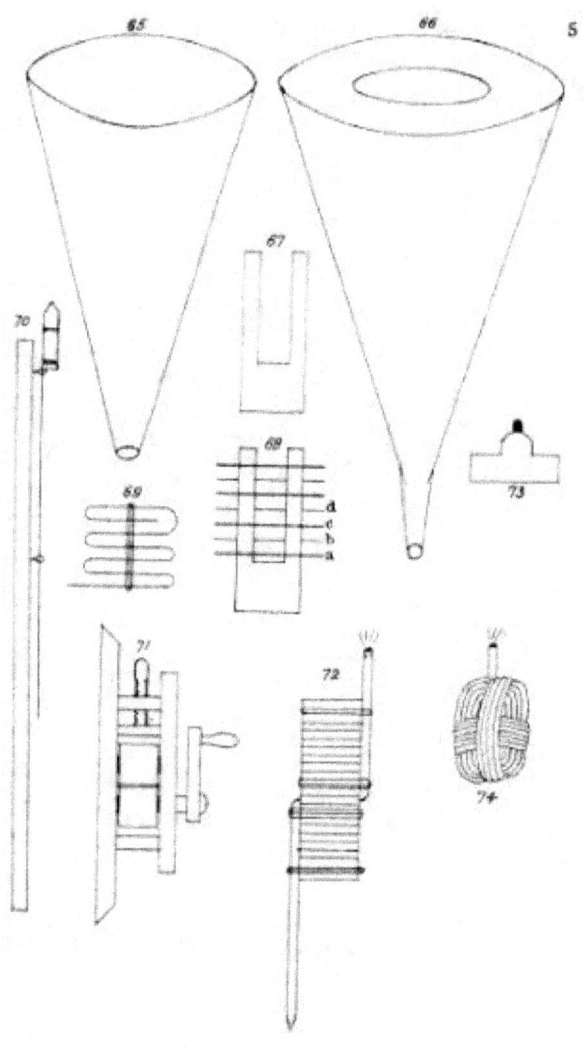

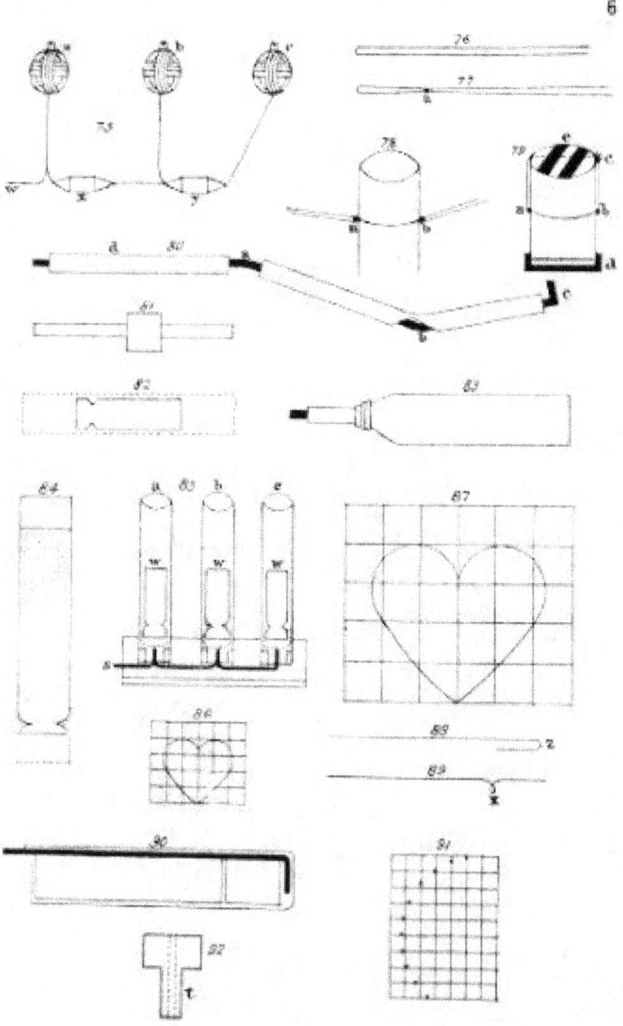

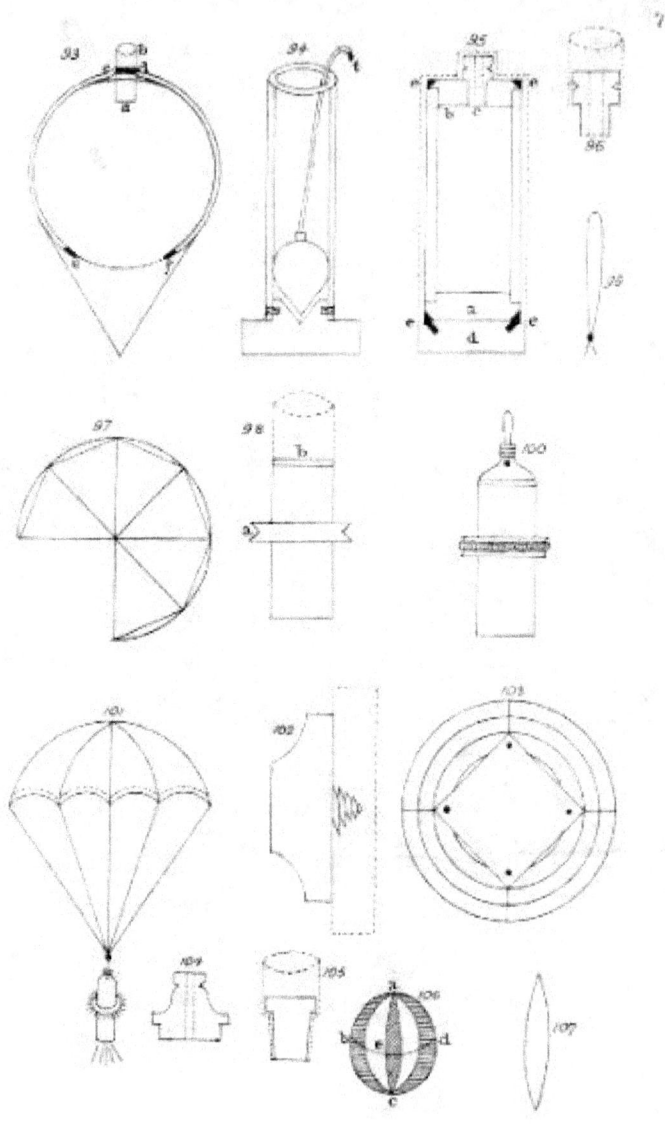

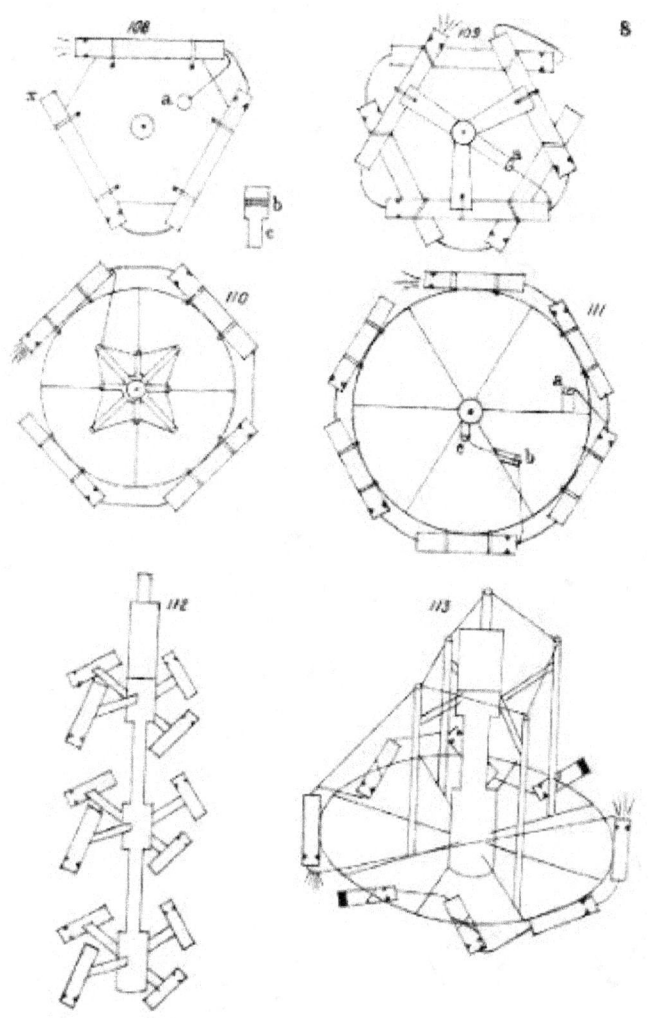

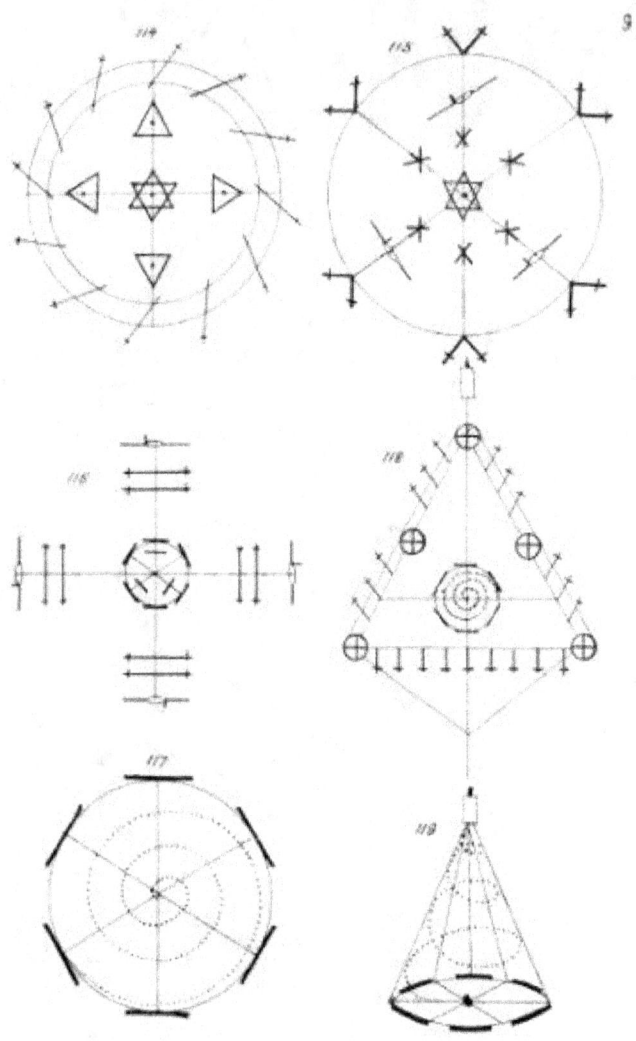

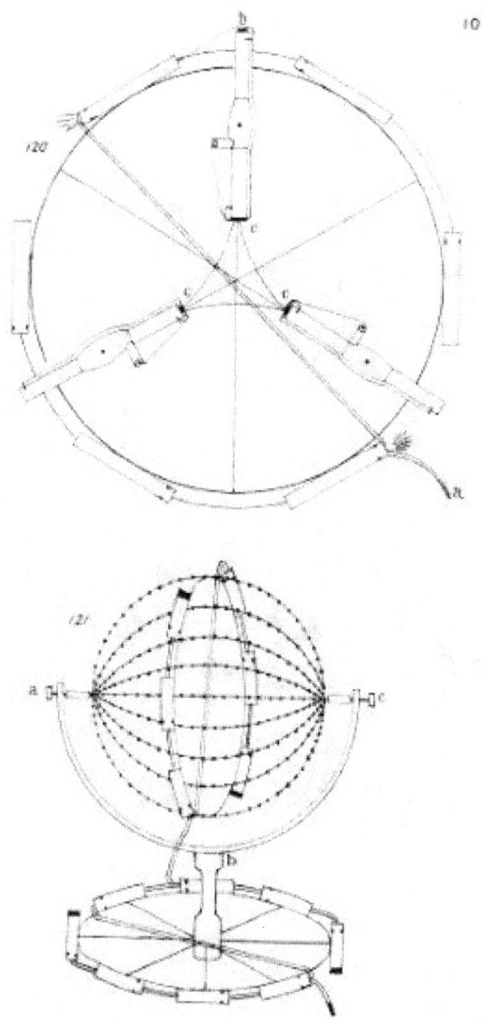

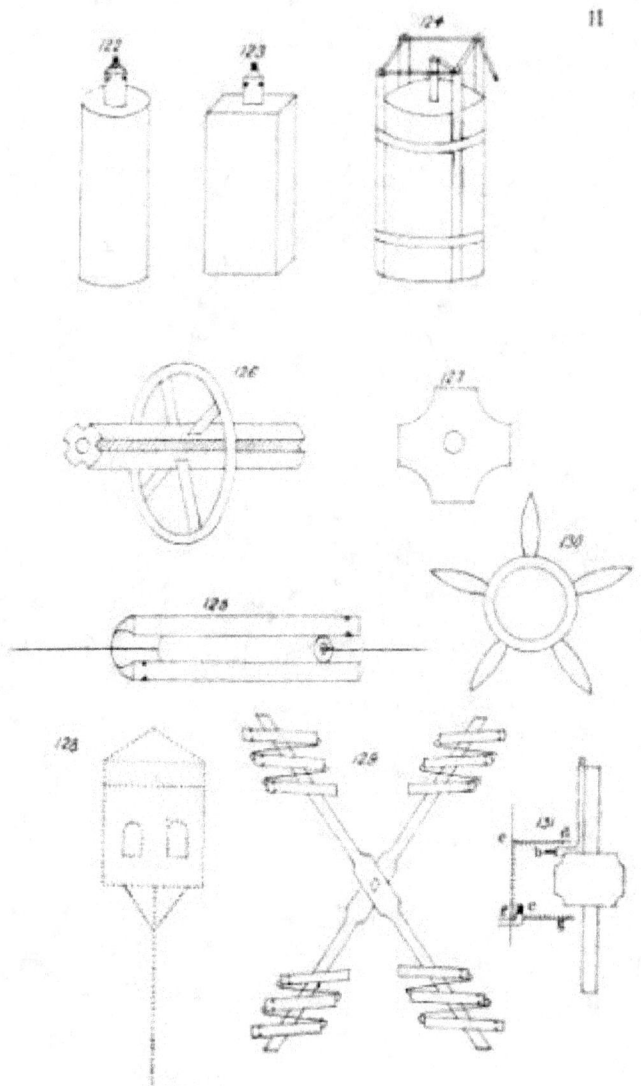

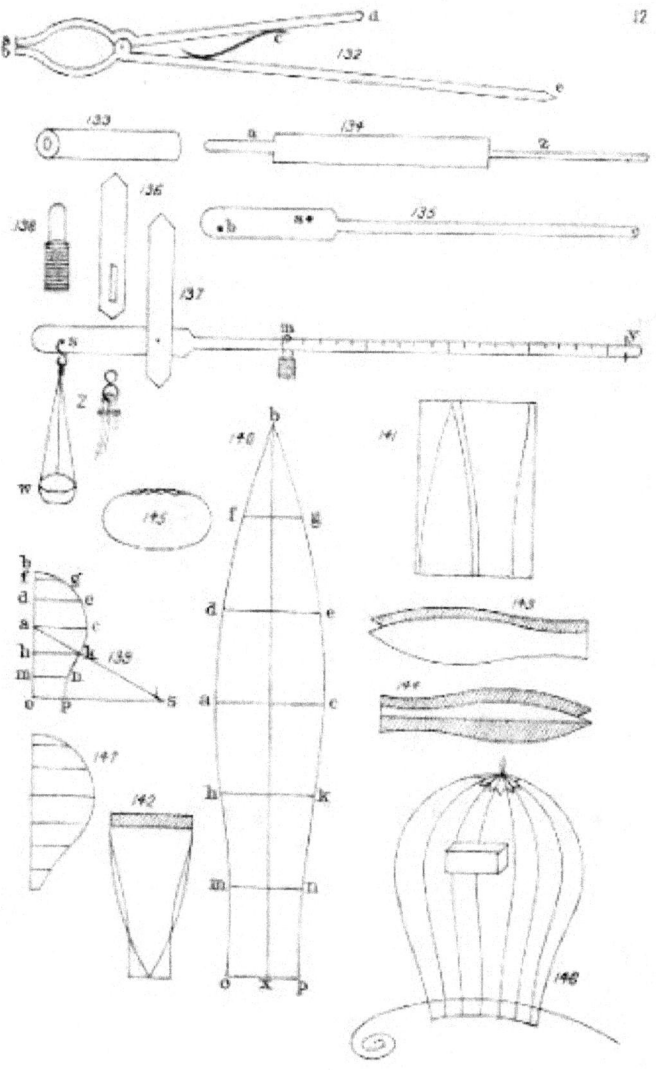

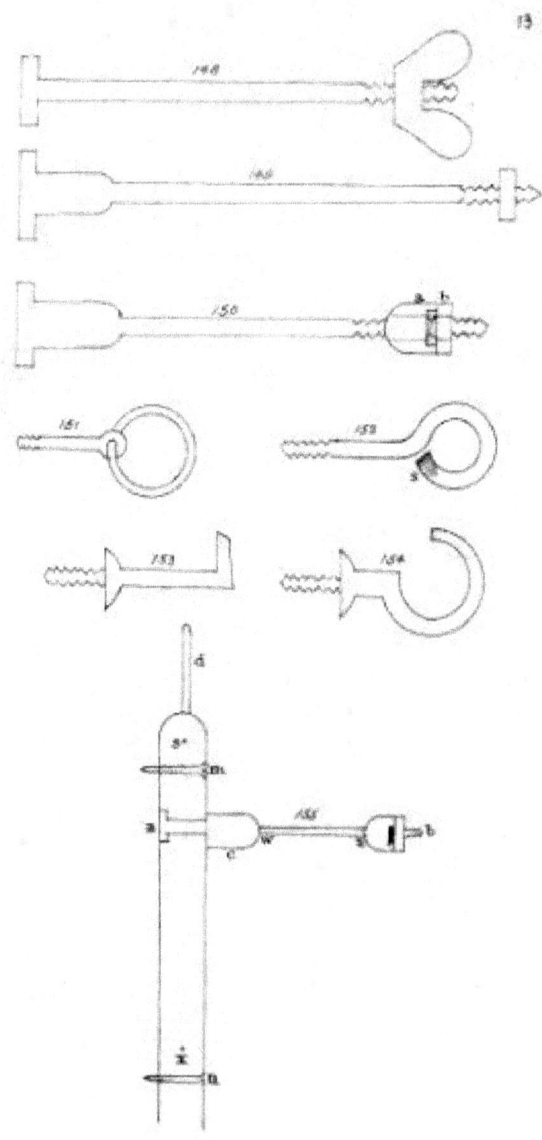

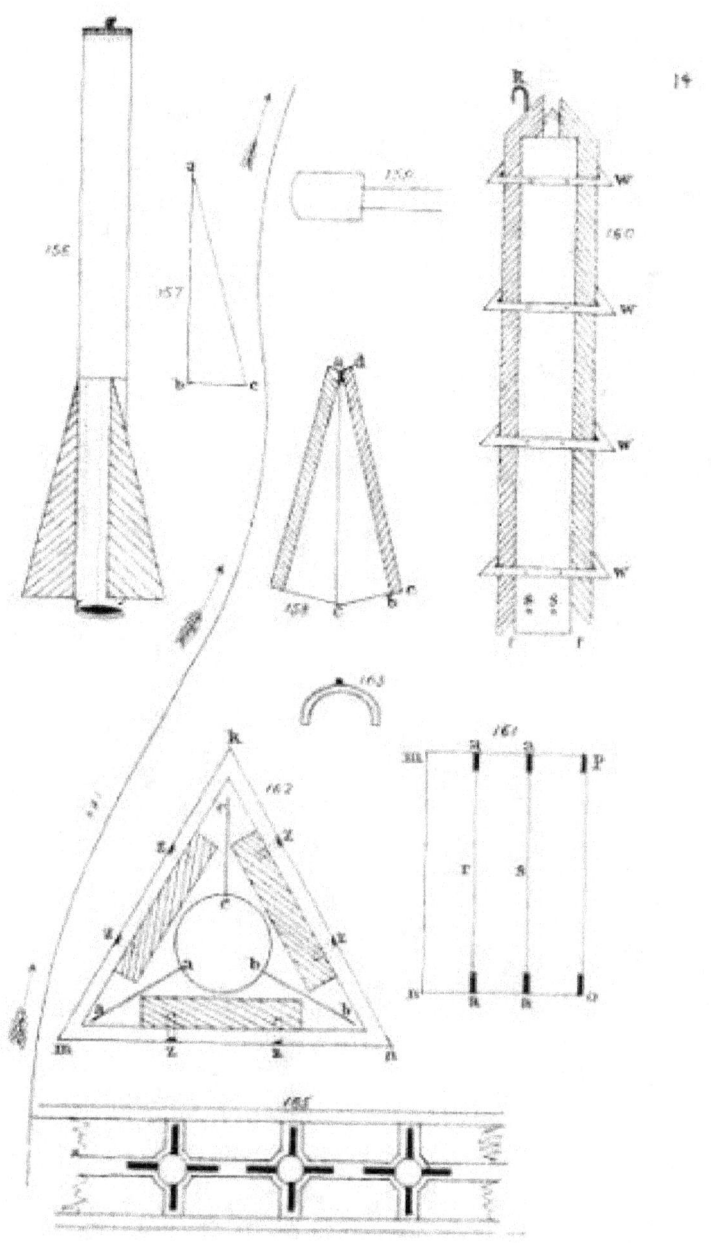

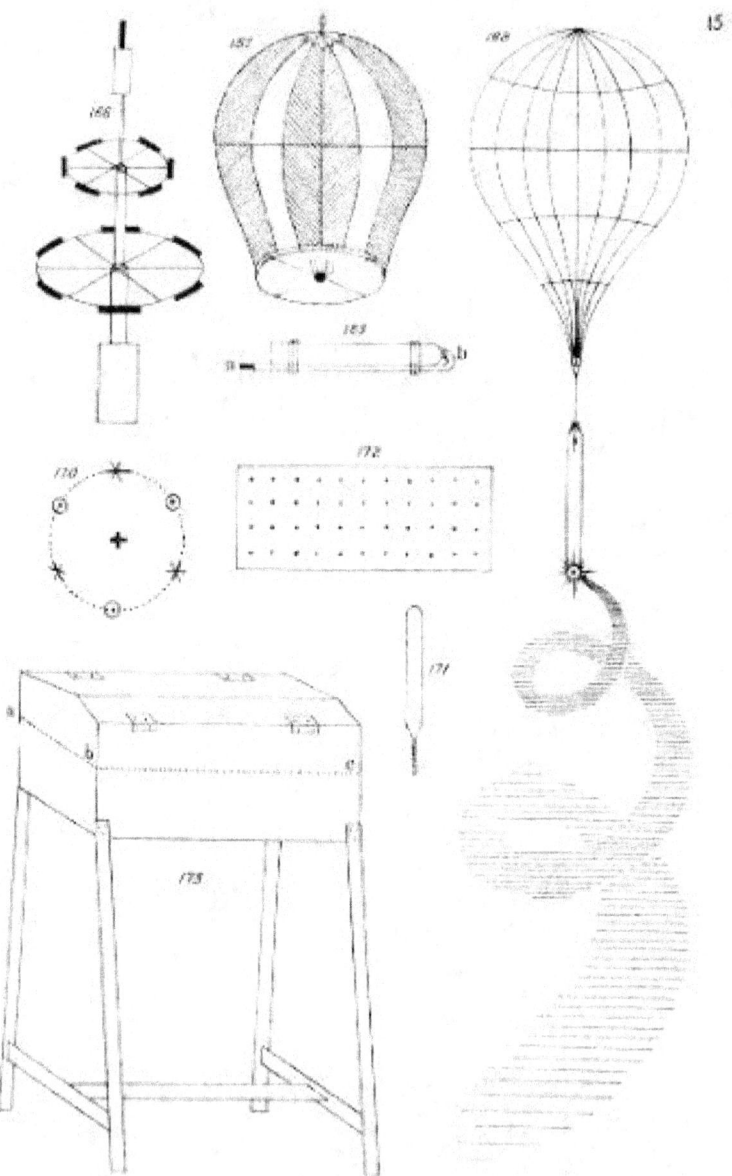

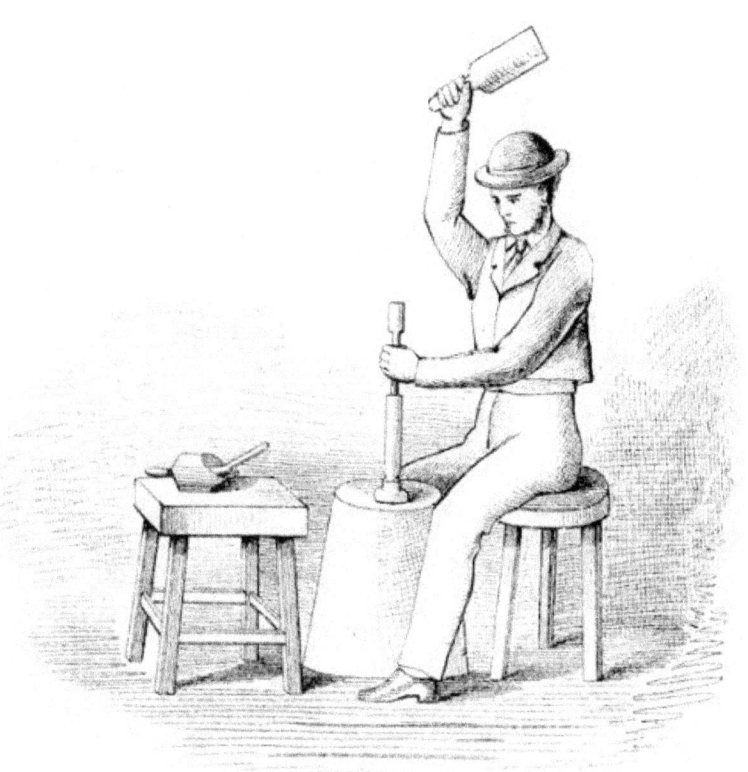

Charging a Case.

www.ingramcontent.com/pod-product-compliance
Lightning Source LLC
Chambersburg PA
CBHW081151180526
45170CB00006B/2022